P9-DBW-959

Hindsight

Justin
Timberlake

Hindsight
& All the Things I Can't See in Front of Me

Justin Timberlake
with Sandra Bark

An Imprint of HarperCollinsPublishers

To Lynn,
who saw what I could be and has
loved me through it all.

To Jess,
who sees who I am and continues to
love me through it all.

To Silas,
who inspires me to see all the love,
all the time, through it all.

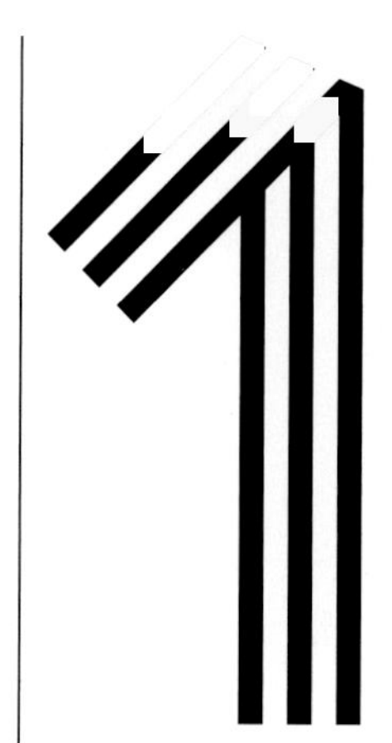

Contents

These

These

These

These

are my memories.

are the thoughts
I have in the middle
of the night.

are the images
I see as I close my
eyes to go to sleep.

are the stories
I will tell my son
as he grows.

There is my childhood in Memphis, and that richness. There are the things I would have liked to have asked my grandfather. There is the music and the movies. The moments when an idea for a song just hits, when I put myself on the line, when I take a creative risk. There is my family, the one I came from, and the one I am creating. And there is the open green of the fairway, because that's how I spend my free time, playing golf, the times I just go off and hit balls for an hour and a half.

I called this book *Hindsight* because it is a long look backward and a deep look inward into everything I am, and how those things connect to where I came from, who I am now, and who I hope to become. It's how it feels when I'm onstage, and you're seeing me, and I'm seeing you. It's wanting to be more present, more soulful, and more humble. It's the ways being a father changed things, and wondering how my son will see me and see himself as he grows.

Ten years ago, I couldn't have written this book. Ten years ago, everything was about forward movement. About taking risks. About trying new things. I didn't look behind me. I didn't care about what was behind me. I cared only about what lay ahead.

Now I'm older. I've made more music. I'm a father. I'm still doing new things, taking creative risks. I have space within myself to look back, to see where I was: to understand what I absorbed when I was a child forming my earliest impressions about the world; to appreciate what I was learning as I became a man; and to reflect on the beliefs and values I am bringing with me as I move into the future.

What I understand now is that there isn't just one thing that I am. There isn't just one thing that I will become.

Every time I make an album, I always want it to sound at least slightly different from the last. I want to explore different genres, and I want to explore different parts of myself. What you hear in my music may not be what you would expect from me, and it may not even be what I ever expected to create myself. But when you hear it, you'll get it; just like when I felt it, I got it.

That's what happens when you forget about labels and look into yourself and make music you love. That was something I had to learn.

The Connection

“The mystery of loving is God’s sweetest secret.”

Jalāl al-Din Rumi
from “Desire and the Importance of Failing,”
thirteenth century

Connections are all around us, and they are inside of us. They inspire and they illuminate. They show us who we are and who we want to be. That's why we make art and that's why we go see it. When we watch, when we listen, we're not getting away from the world. We're actually digging in. You can't help the fact that a song sounds like something you heard when you were a kid, or that you like it and you want more of it. You can't help the fact that a character in a movie reminds you of yourself, or one of your best friends. Our senses and our memories are entangled. That's why I love entertainment, in all forms: it's connection disguised as escapism. That's why we listen to music. That's why we go to a movie, a play, or an art gallery. These experiences touch something within us, sending us back, validating who we are, and showing us something true about ourselves.

Roots

WHEN I THINK OF MEMPHIS, I think of the people and the food and the heat and the humidity. Mostly, I think of the music.

Memphis isn't an easy place to describe. It's a city that is not quite like any other. It's the capital of the blues. It is its own world, with its own way of life and its own sound, and that sound holds so many sounds within in it, just like Memphis holds so many kinds of people in it. Those are the sounds that shaped me and shaped my music.

There are different sides to Memphis.

It's very Southern, and there's not a lot of money there compared to other big cities in the United States, but there's a lot of local pride. That's what I'm singing about in "Livin' Off the Land," off *Man of the Woods*, the album I was writing and recording while I was writing this book.

We lived just outside the city, in Shelby Forest, where it was a little more rural. There was a creek and some woods behind my best friend, Trace's, house, and we'd run around and play back there. Or everybody would get together and play basketball in someone's driveway. There were a lot of girls on my block, and they all had blonde hair, except for one. She was three years older than me, and she wore Jordans, and she played basketball. I was obsessed with her. That's one part of Memphis.

If you're from Memphis, you're proud to be from Memphis. I know I am. Growing up that way was a positive experience for me. Memphis is a part of me and a part of my family and a part of my music. It was the backdrop for all the sounds and sights around me when I was a child, learning what the world was, figuring out what I could be. Who I could be. And the music was all around me.

The first time I held a guitar, I was very young. I remember my grandfather, my mom's dad, putting his Gibson in my hand and teaching me a G chord.

In Tennessee, acoustic was everything.

***

Stax Records was in Memphis, in an old movie theater on East McLemore Avenue. That label established a sound for Memphis, a particular kind of soul that is so intrinsic to my city. Hi Records was also in Memphis, and they put out Al Green, who was one of the voices that really influenced me as a kid. Another old movie theater on South Lauderdale Street became the Royal Recording Studio, where Al Green recorded "Let's Stay Together."

I loved listening to songs like "Let's Stay Together" and "Love and Happiness." When I found out that guy lived down the street, you can imagine the look on my face— "Wait, where? We've driven by his house!" Because you didn't have any idea he lived there when you went by.

Years later, in 2003, I went back home to Memphis to film *Justin Timberlake Down Home in Memphis* (yes, it's a super creative title), a concert special at the New Daisy Theatre, a famous club on Beale Street. When the producers asked me who I would choose if I could pick anyone I wanted to come and sing with me, the answer was obvious: the idol who lived near me when I was growing up, the guy whose truck I used to see parked outside the general store, one of the musicians whose voice helped define the sound of Memphis: Reverend Al Green.

When Al arrived at sound check the day of the performance, he was so joyful, like he was as happy as I was to be there. I thought to myself, "Wow, Al Green likes my music, too? Or maybe he loves his hometown as much as I do." Either way, he stepped forward onto the stage later that night with me, and we performed "Let's Stay Together" as a duet. But, really, it was a group performance because everyone in the room was singing along with us. It was a truly gratifying moment for me as a solo musician. There I was, twenty-two years old, and one of my idols was tipping his cap to me. I couldn't wipe the smile off my face.

That's how I think of Memphis.

SHELBY FOREST GENERAL STORE
LOCKE
WE SELL
FISHING BAIT
& TACKLE
DELI
OPEN

If you're from

you're proud
to be from

PHIS

I was in the car with my mom and her oldest brother, Denny, and an Eagles song started playing on the radio. I began singing in the back seat, and my uncle noticed it right away. That's what they tell me, anyway. Don't hold me to it. I was two. How was I supposed to remember?

As the story goes, he pulled the car over and said, "Do you hear that? Your son is singing harmony with Don Henley."

Apparently, this was very unusual, especially for a little kid. I was tracking the note I was hearing and adding a third, creating something that felt and sounded good to me, but I wasn't aware that I was doing anything at all. Harmonizing was part of the way I interacted with the world I heard around me. That's how I've always heard music.

I never meant to be a singer. Singing was just something I could do. I could always hear the notes. I could always harmonize. My relationship with music was intuitive and completely natural to me. It's like the way you don't think about seeing when you look at colors. You just see them. That's what music is like for me. When I hear one note, I can hear the other notes in the scale that and me, with my toy banjo, standing in between them. We put the image on a T-shirt for the *Legends of the Summer* tour I did with Jay-Z in 2013. That tour was groundbreaking for me. Being outside and playing stadiums was something I had experienced before with 'N Sync. But that summer, the air reminded me of those hot nights when I was that little boy just watching his daddy play and sing.

I loved standing in the front row. One of my mom's favorite memories of that early time was watching my reaction when a little girl decided to stand in front of me at one of their shows. Apparently, I took my banjo and walked right in front of her, backing her up like I was Charles Barkley protecting the hoop, never losing sight of the band. My mom also says I almost knocked the poor girl over. She's been known to exaggerate, at least when it comes to stories about me, but if I'm being honest with myself, this whole story could very well have happened exactly as she described it, because it's always been true that as close as I could be to the music, that's where I wanted to be.

FIRST
TENNESSEE
BANK

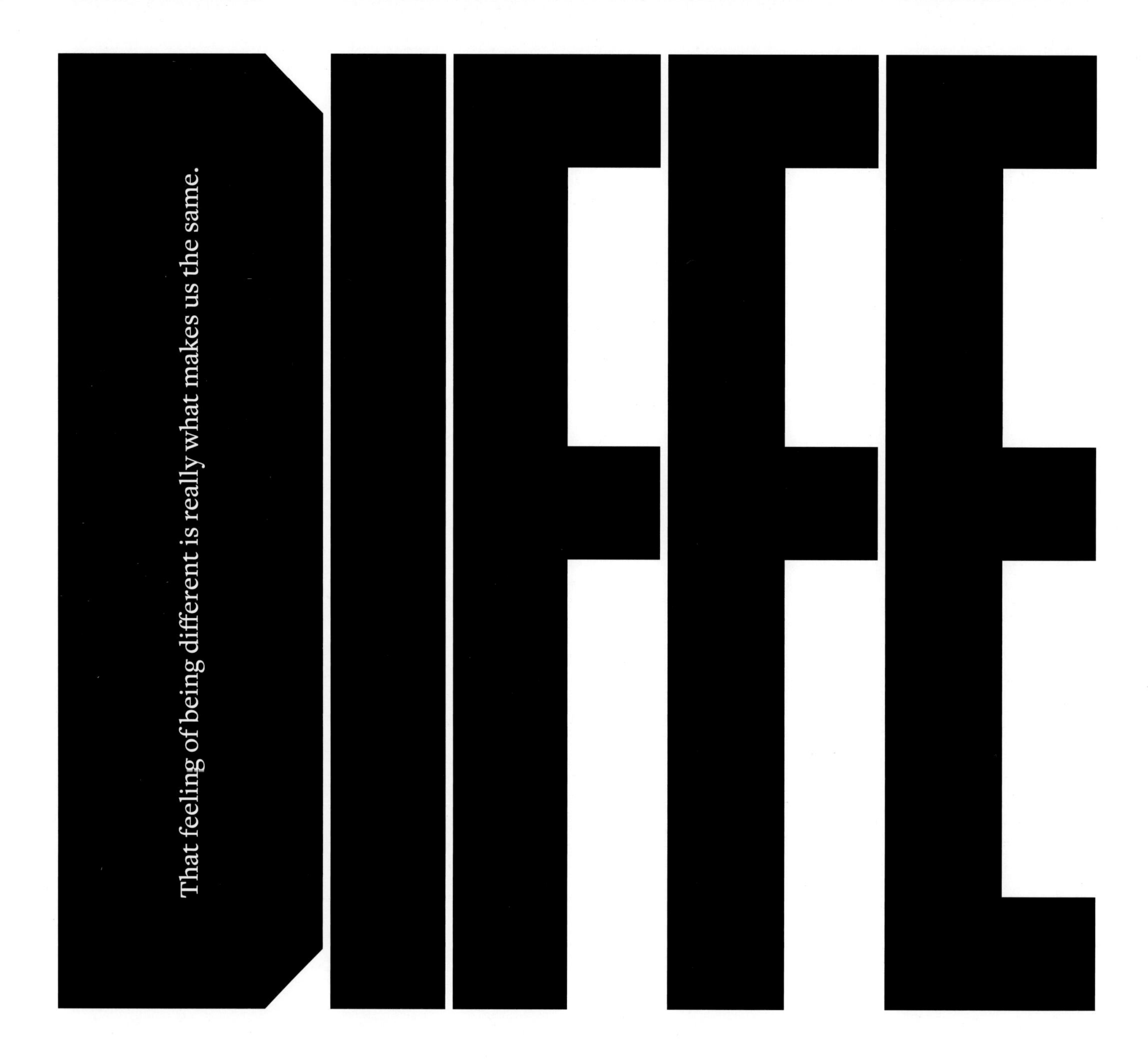

You Don't Have to Be Related to Relate

MY PARENTS DIVORCED WHEN I WAS TWO, AND WHEN MY MOM remarried, I lived with her and my stepdad. I was the only child.

I remember having a lot of fun in my neighborhood, but at school, I had a hard time relating to the other kids. Music was my favorite class, and when I was in middle school, the music program got cut, like arts programs in a lot of schools often do. This was a small school in a small town that didn't have a budget for things like that. I wanted more, but there wasn't more to have.

Now I realize that I probably needed more information and more stimulation than what was available to me. As a young person, I didn't have the awareness to identify what was wrong, and the adults around me didn't know, either. I dealt with it by spending a lot of time alone, and that time revolved around music. After school, I would go home and put on a cassette and absorb the songs that poured into my room. Music was my refuge. The way I felt when I listened to music was better than the way I felt all day.

The older I get the more I understand that so many people live in circumstances they can't control, or in places that just don't feel right to them. That feeling of being different is really what makes us the same. We have our own struggles, yet we want the same things. We want human connection, a place to feel at home, and pizza.

Even if people seem guarded or bashful, more than anything, they want to relate.

That's all I've ever wanted to do.

***

Every Sunday, I would go with my family to Shelby Forest Baptist Church, where my dad worked as the choir director. His dad was the preacher, and his mom, my grandmother, was the organist. That church was a part of my family as much as my family was a part of the church.

When I was eight years old, my dad asked me if I wanted to come up and sing a hymn with him in front of the whole congregation. I don't remember what song it was, but I do remember that I sang a harmony to the melody that my father was singing. I could hear the notes in my head, and I knew where they needed to be.

Being up there with my father must have given me the courage I needed, because I wasn't intimidated at all. What I remember most is looking out at all the people. When I saw that they were listening, really listening, that got my attention. It was this moment of pure connection between me and them, and I felt it completely. When I saw what singing could accomplish, when I realized that this was a way for me to connect with people, I started liking the idea of singing a whole lot.

CHURCH IS THE BEST PLACE
TO GET YOUR START,
LIKE WE SAY IN THE SOUTH,
BECAUSE EVEN IF IT'S TERRIBLE,
WHEN IT'S OVER
THE CONGREGATION STILL SAYS

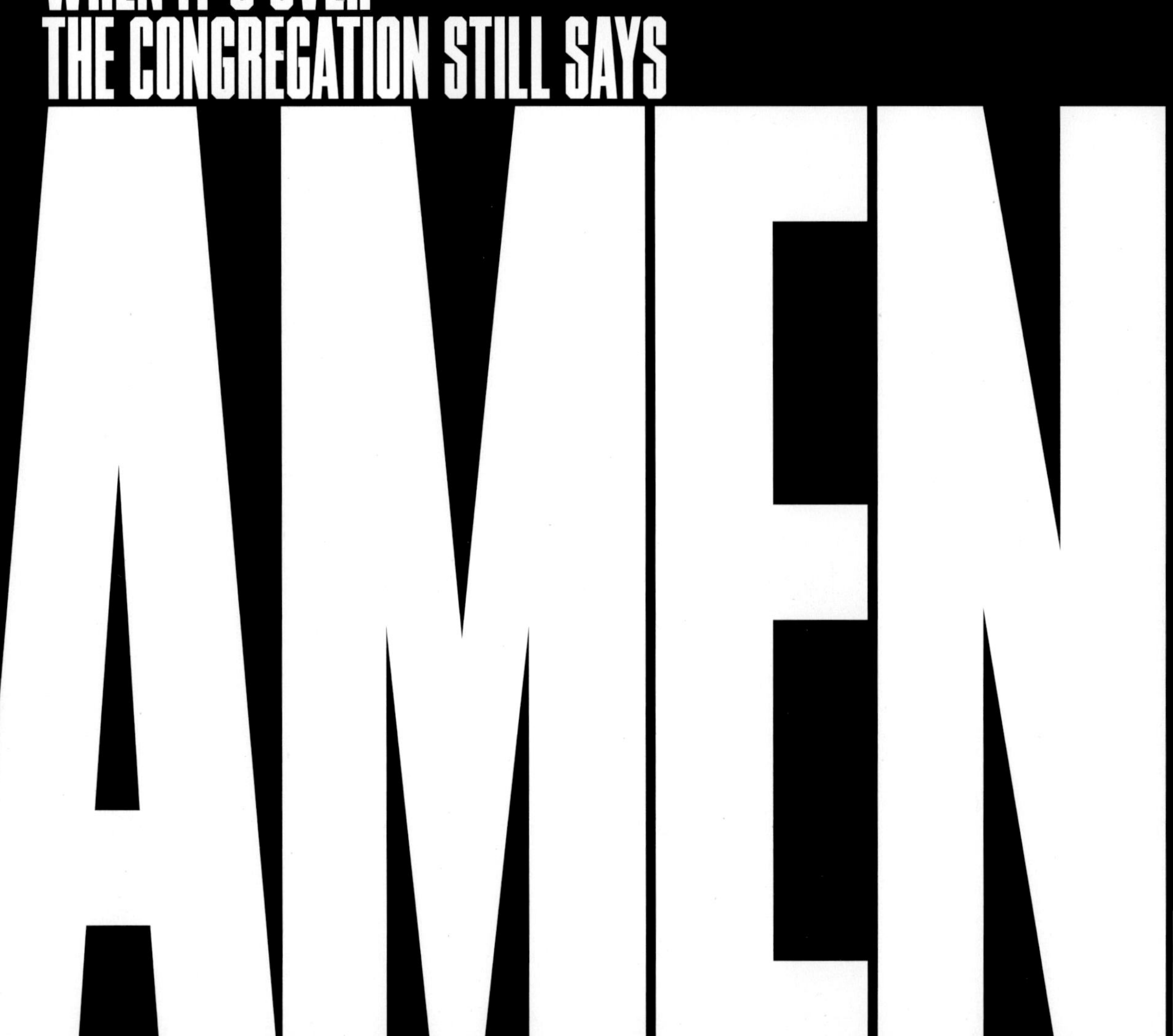

A Night at the Opera

HEN I STARTED LISTENING TO MUSIC, it was all cassettes. My father had a vinyl collection and a record player at his house, but I had never really played a record on a record player until I was nine. I used to spend every other weekend with my dad, and one night—maybe the album was off the shelf already—I got my hands on Queen's *A Night at the Opera*. I could not stop looking at the album artwork—it had this scene like a mural, with a swan and some other animals. I was so intrigued with those visuals. They made me want to hear whatever a picture like that could sound like. I put the record in the player, set the needle, and turned it on. I started listening to the whole album, song by song. And then "Bohemian Rhapsody" came on.

I listened to the song, and as soon as it was over, I needed to hear it again. I locked myself in the room and turned the blinds so the space would be completely dark, and then I turned off the light. I listened to the song the whole way through in the dark. It blew my mind.

I turned the lights back on and adjusted the needle, and let it go again. I probably listened to the song ten times in a row. The harmonies made sense to me. Freddie Mercury's vocal arrangement made sense to me. All the parts made sense to me. That moment when I discovered this incredible music and my deep connection with it is a wonderful memory.

I walked around singing "Bohemian Rhapsody" for the rest of the weekend, and my dad would chime in, singing the lower parts. "No, no, no, no, no, no, no!" Messing around together, having fun, singing this song—that's the connective power of music. That's what I love about it.

A Night At The Opera

agnificooooooooooooooooooo

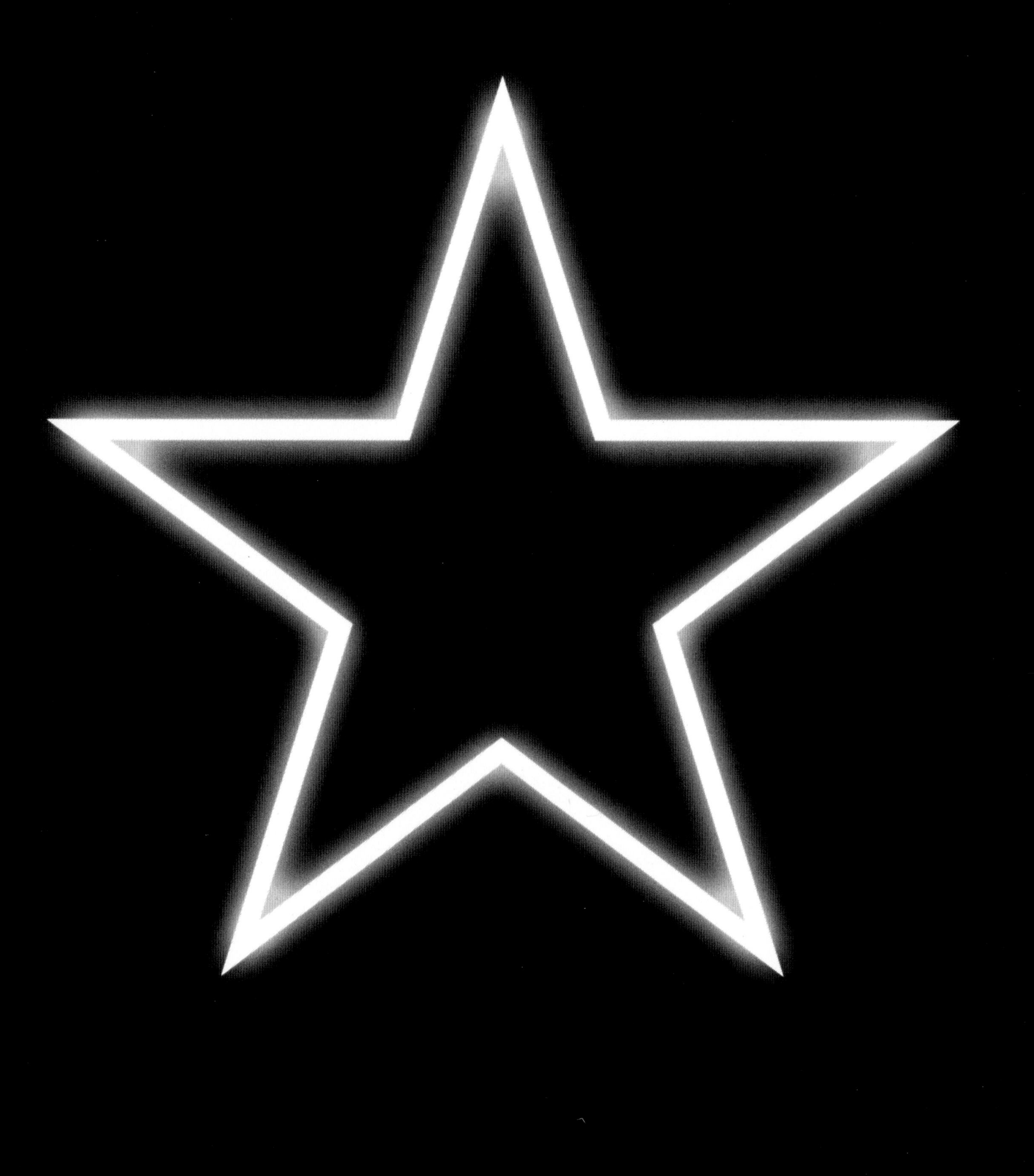

A Free Trip to Disney

I HAD TWO FRIENDS WHO LIVED ACROSS THE STREET FROM ME, Amber and April. Every day when we got off the bus, I would go straight over to their house and we would watch Kids Incorporated and the Disney Channel's *The Mickey Mouse Club*.

I loved watching *MMC*. All I had ever wanted to be was Sam Malone, to be as cool as Ted Danson in *Cheers*. He was the coolest. But the kids who starred on *MMC* were close to the same age as me. I felt like I related to them, like I knew them better than I knew 90 percent of the kids at my school. Not only were there twenty oddball kids who were on the show, there were also four hundred kids in the audience watching. I loved that—a live studio audience. It was almost like a kids' *Saturday Night Live*.

Seeing that there were kids in the world who got to be on TV, singing and dancing and doing sketches, made me wonder how they got to do that. I knew I could do what they were doing, and I wanted to.

So when the producers for *Star Search* came to town and held auditions at the Oak Court Mall in Memphis when I was eleven, I went. *Star Search* was the biggest talent show on TV in 1992; it was *American Idol* before *American Idol*. I auditioned twice. The first audition, I sang Garth Brooks's "Two of a Kind, Workin' on a Full House," because he was one of the biggest country singers in the world and because I loved singing country music. When the local haunt, Casper Creek, would have teenage night, parents could bring the kids, and we'd go. Everybody would do country line dancing, and I would get up with the band and sing a song. They knew every country song, and I knew every country song. I had a knack for being a belter, but I would have to ask them, "Can you play it in a key higher?"

For the second audition, I sang Percy Sledge's "When a Man Loves a Woman," which was all over the radio because Michael Bolton recorded his own cover of it. He was a pretty big deal, too.

The producers chose me. It was incredible.

Star Search was filmed in Orlando. I remember my mom saying, "Well, at least we get a free trip to Disney World." That was one of most exciting things about the whole scenario.

We took a road trip to Florida to do the show. When we got there, we met people from all different walks of life on the set, like Dave Chappelle. He was there because *Star Search* had a comedy category. That was the first time I met him, so there's that weird fact. And hey, look, we both did all right for ourselves!

The producers were very clear on what they wanted me to sing: country music, not soul. That decision taught me a valuable lesson at a very young age about how television works. When you're at a talent show, you choose a song that you can really sing, the song that makes you sound the best. When you're on television, the producers choose the song that will round out the program, that will create the most amount of diversity or whatever they think people at home might want to watch.

They had me sing an Alan Jackson song, "Love's Got a Hold on You." I wore a cowboy hat and boots. It's a great song, awesomely written, but it didn't require a lot of vocal range. I was a little bummed out about being told what to sing, but don't get me wrong, I was really glad to be there, to have this experience of being on the biggest talent show in the world.

I didn't win, but some amazing things happened as a result. What I hadn't known when we came to Orlando was that *Star Search* was filmed at Disney-MGM Studios, on a soundstage in the back lot. And right next door, across a hall that split the two, was the soundstage for the set for the Disney Channel's *The Mickey Mouse Club*.

While we were waiting to rehearse for *Star Search*, we'd work with the tutors who were there to give us our school lessons. Sometimes we'd ask if we could sneak a peek at the *MMC* set, and they'd walk us in. The show wouldn't be filming, and the lights would be down on the soundstage, but I could see where the set was and where the audience would sit. Looking in, literally seeing the set that I watched on TV with my friends from school, was such a big moment for me. It made it all feel real, feel possible, feel like something I could really be a part of.

Thank You, *Star Search*

THE WAY HOME, ROAD-TRIPPING back to Memphis, we pulled into a dingy, dodgy motel—the kind of place where you could get only *Perry Mason* on the black-and-white TV—somewhere in Tennessee.

While we were watching, an ad popped up that said, "Local call auditions for the Disney Channel's *Mickey Mouse Club*."

I looked at my mom, and she looked at me.

My mom is a big believer in kismet, in fate. For her, this was a sign.

She said, "Do you want to go audition?"

I said, "Yeah, why not?"

We went to the audition the next day. The place was basically an open ballroom, filled with all these kids wearing smiles that looked plastered on. I knew that I was a weird kid, but these kids were really weird. I didn't want to wear a smile like that. I just wanted to be me.

The people from casting explained how it would all work. They wanted to see if we were funny, so they gave us a sketch-comedy monologue, and told us to study it. I had never really studied a script, but I understood what I was supposed to do. They wanted to see if we could dance; that part was freestyle. They would play a song and we would show them our moves. And they wanted to know if we could sing. That was the best part, because I could sing any song I wanted.

When it was my turn, I went into the audition room by myself. They didn't let parents come in with you, so it was just three adults and me. One of the adults was the casting director, Matt Casella.

I did the sketch. I danced. I sang "When a Man Loves a Woman," and even at such a young age found irony in the fact that it was the song I had wanted to sing on national television a week prior.

After I sang, Matt said, "Wow."

That was when I knew I had the room.

Then they asked me a question. "Why do you want to be on *The Mickey Mouse Club*?"

That was a pageant question—and when I was eight, I had done some pageants. I had wanted to be in a talent competition for a while, but the only local outlet for that kind of energy was pageants. You could only do the talent competition if you participated in the whole pageant, though, so I did it. All the girls and me. I got a lot of shit for that, but I also got some great experience. Because you always have to answer that pageant question.

I said, very drily, "I'm in it for the money."

That got a huge laugh from Matt. I don't know why I knew that would be funny to an adult. But it was.

We went home, and I got the callback. My mom and I looked at each other, and I said, "Hey, it's another free trip to Disney World."

Later on, we would come to learn that I was one of roughly twenty thousand kids who had auditioned all over North America. They had chosen twenty-seven of us to go to audition camp.

Audition camp was held down in Orlando, right next to the *Star Search* soundstage and Disney-MGM backlot. Oh, the irony. *I was just here. And here I am again.*

All the kids who got callbacks spent three days working with a choreographer, a vocal coach, and an acting coach, to prepare for a bigger on-tape audition. Twenty-five years later I learned that Jeffrey Katzenberg, who worked at Disney at the time, had seen that audition, which I found pretty funny, because when we worked together on *Shrek the Third* and on *Trolls*, he never mentioned it.

After audition camp, we went back home and waited. Radio silence for two weeks. And then I got a phone call from Matt Casella.

He said, "We'd like you to be on our show."

Then it was me and my mom and my stepdad jumping around the house with unabashed excitement. I didn't know what to expect when we got that phone call, but I knew I had done my best at the audition camp because I wanted to be chosen. My reaction was definitely a mixture of emotions. I knew that I had a knack for this thing, but it was hard to believe that it could really be happening. I was just a kid in a small town in Tennessee who watches *The Mickey Mouse Club* every day after school, and now I was going to be on this show. It was pretty overwhelming.

That's why I tell younger artists to dream big. Dream as big as you can. If you can see it, you can make things happen for yourself.

How did I end up on *The Mickey Mouse Club*? Because of *Star Search*.

Thank you, *Star Search*.

I'M IN IT FOR THE MONEY.

M-I-C

NOW SHOWING
MOUSE CLUB'
NOW SHOWING

M-I-C (See You Real Soon)

IMAGINE THIS WEIRD MAGICAL PLACE WHERE TWENTY OR so young people, ranging from twelve to eighteen years old, starred in a show that was performed in front of a live studio audience, with cameras in front of them: that's *The Mickey Mouse Club*.

We did sketch comedy and performed our own renditions of songs that were popular on the radio at the time. I had the opportunity to learn how to engage people from a very young age, how to get them to connect with whatever I was doing onstage by watching them and learning from their reactions. I got to see how to play to the camera and move people at the same time. I loved being on that show. It was fulfilling in so many ways, and so much fun. Matt Casella had also cast Ryan Gosling, Britney Spears, Christina Aguilera, Keri Russell, and JC Chasez—he was the one who found all of us. We had a very good time on set—and off.

We had employee cards that gave us access to the theme parks for free. Ryan and I once stole a golf cart and drove it to the employee entrance for the Tower of Terror. We went on that ride twelve times in a row.

K-E-Y

We got reprimanded for that. It was worth it.

A Free
Trip to
Europe

FTER I WAS ON the show for two seasons, *MMC* was canceled, and I went back to Tennessee. I wanted something more, and I finally got it, but then I had to go home. It was like a switch had been turned on inside me, and then I had to turn it off.

I was thirteen years old, and I had so much energy, but with the show over, I had nowhere to put it. So I joined the basketball team and all the clubs that interested me at school. I came home a lot more sophisticated and aware than I had been before the show, but I tried to downplay it because I just wanted to seem like everyone else. I knew if I started talking music or singing, I'd be called a sissy and sometimes worse than that. Kids can be so mean to anyone who seems different. I always felt very aware of that, and to divert the attention, I developed the habit of using humor to disarm people in order to endear myself to the kids around me. I became the class clown, disrupting class with my bits, not caring if the teachers were mad, only wanting to be accepted by the other kids. I started getting in trouble. I smoked pot for the first time. I got myself a can of tobacco and almost got expelled for that.

Not everybody understood what I was going through, but my mother did. I give her a lot of credit for that. My teachers called her in for a parent-teacher conference and told her that I was doing too much. At that point, I was president of the student council, point guard on the basketball team, and running the Beta Club.

My teachers thought I should quit some of those things.

My mom said, "I don't think he's doing too much. I think he's bored."

She knew that I needed more.

My mother had a conversation with Robin Wiley, our vocal coach on *MMC*. Robin was a songwriter who lived in Nashville. She told my mom that I had a gift and that she wanted to help me nurture it, so we started doing these trips out to Nashville to learn a little bit about songwriting.

JC was also working with her. He was older than me by a few years, and he picked up the knack for songwriting a lot quicker. But we were both strong singers at the time.

Then my mom got a phone call from a guy named Chris Kirkpatrick, who said that he was starting a vocal group. He wanted to know if I wanted to go down to Orlando for a meeting.

Immediately, my mom and I looked at each other. "Sounds like another free trip to Disney."

So for what felt like the eightieth time, we went back to Orlando to meet Chris and Lou Pearlman, a record producer and manager described to us as "a guy who would put up the money to start a group and get them a record deal." Unfortunately, years later, he would go on to be thrown in jail for twenty-five to life for embezzlement. That was another valuable lesson that you don't understand without hindsight. I will leave it at that.

JC and I went down to Florida at the same time. We were there for a week. We met Joey Fatone then; coincidentally, Joey and Chris both worked at different shows at the Universal Studios theme park, so they already knew each other. Robin came along too, to help us figure out how to put a group together. She worked with us to see what everybody's range was and where they would sound the best.

But none of us could sing bass.

We called Bob Westbrook, who was my first vocal coach. He said, "I know exactly who could do it—but his mother will never let him."

We said, "What's his name?"

"Lance Bass."

We couldn't stop laughing about his last name. . . .

We convinced Lance and his mom to come to Orlando and meet with us. Somehow, my mom said whatever Lance's mom needed to hear.

That was the beginning of 'N Sync. At that time, Nirvana was god. Pearl Jam was king. We were big fans of all types of music, but we wanted to model ourselves after groups like Boyz II Men and New Edition. There was no space for five white boys who wanted to dance and sing, but we had heard about this group called the Backstreet Boys that Lou Pearlman was involved with before us who became enormous in Germany. We busted our asses to do a showcase that we videotaped and sent to record labels, and within a few months, we got a deal with BMG Munich.

BMG sent us to Europe to build a story, because pop music was just so big there, and bring it back to America. My mom was traveling with us as a guardian because Lance and I were underage. We were playing festivals with sixty thousand people; when we would finish a show and go to the bus, a thousand girls would be running behind us, falling over one another and screaming.

It was huge.

We had a lot of fun, and we really cared about what we were doing. We wanted to be good at it.

There were also some very weird parts of that era as well. One time we got booked for the wrong festival somewhere in Scandinavia. The festival consisted of a bunch of heavy metal acts and "this up-and-coming boy group from the USA." We spent our twenty-five-minute set dodging glass bottles being thrown at us by tall, blond, heavily tattooed people.

In Europe, we had our hard knocks and our bigger-than-life moments. I got to see the world, and then we came back home. Our big break in America happened with a 1998 Disney Channel special called *'N Sync: Live in Concert* that had a big teen audience.

That was a crazy time.

Bring It on Down

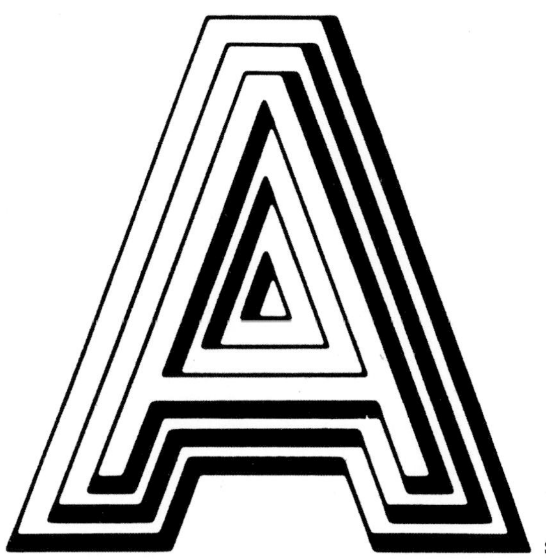

S A KID, singing was the ultimate way for me to connect with people, until I discovered I could genuinely make people laugh. Those times when I had my mom going, "Oh!" and falling apart laughing made me feel incredible, and I wanted more of that feeling. My mother's laughter carbonated me. It made me feel alive. It made me want to try harder to make her laugh more.

Any time my mother and stepfather had friends over in the evening, I'd stay up late and try to make them all laugh. When they would start laughing at something, I would seize upon it. And if I got a laugh, whatever I had been doing—it could be physical comedy, like, a fake trip-and-fall—I would just keep elaborating on it and turn it into a whole bit.

To make my mother laugh like that, to make adults laugh—I knew that if I could do that, then I must be funny. (Looks don't count. Or do they?)

Making people laugh still carbonates me. It all tracks back to being a kid, to the electric charge I got when the adults started laughing.

TEA SALON

CUP
O'
OUP
ARMY
UP O' SOUP

***

The first time I went on *Saturday Night Live*, I was with 'N Sync, and we were the musical guest. Coming back as a host was another thing entirely.

The first time I hosted the show, Lorne Michaels, the show's producer, said to me, "Don't worry, we're going to make you funny. Don't worry." The more you get to know him, the more you know he was being his dark, sarcastic self. But I wasn't worried. Everything I've done on *SNL*, I did on *The Mickey Mouse Club* when I was twelve years old. What I learned back then, I have brought forward with me as an entertainer.

When we were coming up with ideas for skits for the first hosting gig, I knew people would want to see me do physical, musical comedy. Around that time, there was this story going around about Brad Pitt's first job in Hollywood, that he had been the El Pollo Loco chicken. The legend goes that he'd dress up in a chicken costume and stand out in front of El Pollo Loco, dancing around, waving his arms at traffic, flipping his sign and saying, "Come to El Pollo Loco."

We decided to dress me up as a funny character, and we just started riffing. What if we added all the popular rap songs? What if there were a competing mascot, who was just terrible? That was the beginning of the whole mascot sketch. That's how I became an omelet.

That mascot bit became the sketch "Bring It on Down to [Whatever]ville," which we've done every time I've played *Saturday Night Live* since.

When you host a show like *SNL*, you come in bearing your own weapons. With my years onstage as a dancer, and even further back than that with my parents as an audience, I knew that I could use my physicality as a springboard into something funny. When one of the writers, Paula Pell, brought the idea of the dancing mascot to me, I knew all we had to do was to make sure the medley of songs we picked to spoof were fun and recognizable, and I could do the rest. We added those goofy, huge gloves into the mix—and we were set.

There are about eight hundred people in the *SNL* studio audience. We made them laugh; millions of viewers at home heard their laughter and paid attention. Everyone connected with the spirit of joy that was infused in that bit. Especially me.

WEEKENDUPDATEWEEKENDUPD

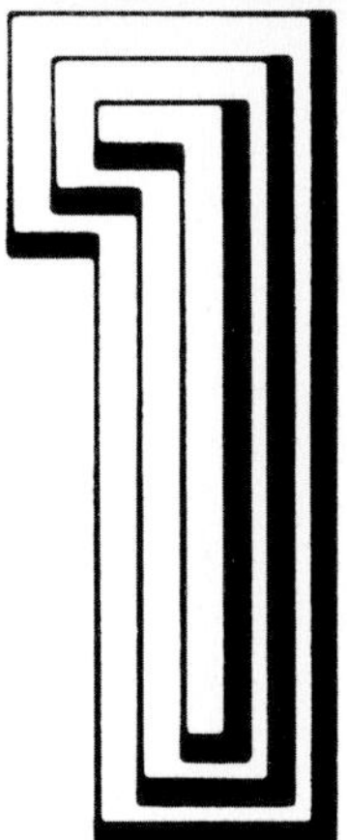

October 11, 2003
Season 29
Episode 2

December 16, 2006
Season 32
Episode 9

May 9, 2009

Season 34
Episode 21

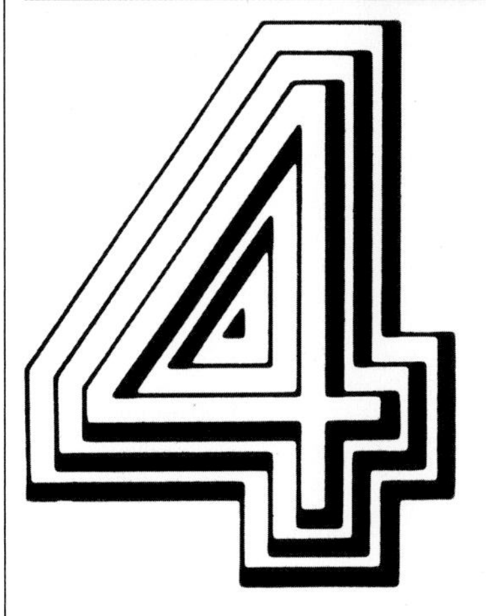

May 21, 2011

Season 36
Episode 22

March 9, 2013

Season 38
Episode 16

THE KENNEDY TAPES

I'M NOT THAT KIND OF GIRL.

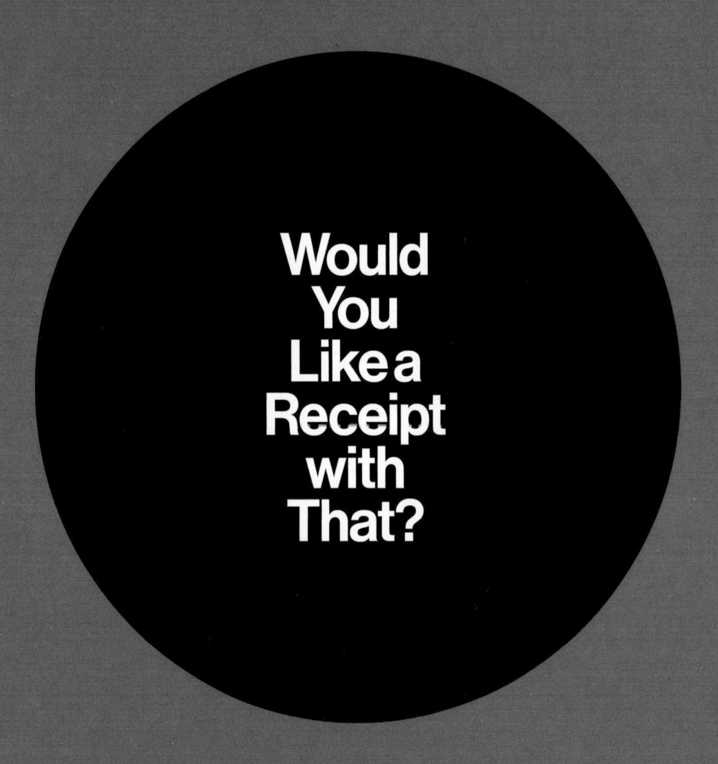
Would
You
Like a
Receipt
with
That?

MAKING SOMEONE LAUGH IS A KIND OF connection. So is laughing together with other people. So is realizing that other people are amused at or inspired by or curious about the same things you are.

I grew up listening to the R & B sound that was so big in the nineties, and so did Andy Samberg, who is close to my age. Around Christmas 2006, we were working on a bit that began with a loose idea about that nineties' guy-group sound, complete with harmonies. We wanted to do a joke about guys who are still stuck in that era who are idiots, but who are also very genuine in terms of the sentiment behind this special gift they are giving.

Yes, I'm talking about "Dick in a Box." Once we had the idea, we just went with it. We came up with the concept on a Thursday and played it on *SNL* two days later, and it immediately became the biggest thing on the internet.

Let me be clear: we were very aware of how inappropriate this sketch was. But sometimes the perfect algorithm mixes itself.

I think I'm most proud of the fact that the following Halloween, grown men all over the world were masquerading in fake beards and silk suits, with a box attached to their appendage, singing "Dick in a Box." It was a number one costume that year.

That's what I love about comedy. When millions of people who don't even know each other are all laughing about the same thing. I think Andy Samberg and I share a certain quality that made that happen. We don't actually care if you are laughing at us. As long as you are laughing.

1

2

3

Cut a hole in a box.

Put your junk in that box.

Make her open the box.

That Thing with the Shoulders

EVERYTHING WE HAVE SEEN AND FELT AND TOUCHED throughout our lives stays with us, infuses us, especially what we experienced as kids. All the music I listened to, all the movies I saw, all the people around me, my parents, my grandparents, their friends, the neighbors who had that dog that never stopped barking, the characters who went to my church—all those people, places, and things are all stored up inside me, waiting to be called on. When I'm trying to create a comedy bit, write a song, or bring something to a character as an actor, I reach in—sometimes subconsciously, sometimes not—and pull out what I need to fill whatever it is that I'm trying to make.

All the characters I do are based on real people. Whenever I'm creating, I channel all my inspirations, all of what I've seen and felt in the world around me and from my inner world, and an image comes to me. All the material I do comes from connecting to my awareness and my experiences. All the silly things, all the sad things—they all come together in these moments where an image emerges, and then I just follow it where it wants to go.

Partly, I drive it. Partly, it drives me.

I did Classic Peg alongside Kristen Wiig, who is a genius, in her classic "Target Lady" sketch on *SNL*. We were in a writing session together and she asked me if I could do an impression of an older Southern woman. I immediately remembered my grandmother's friend. I saw her at our church every Sunday. She had a huge, hearty laugh that shook her whole body.

Kristen asked me, "How do you do that with your shoulders?"

I said, "I don't know. How did she do it with hers?"

I based Classic Peg, very simply, on Bea Arthur's character Dorothy on *The Golden Girls* and my grandmother's friend, who always had this certain expression on her face, this funny way of conveying a story. That all became a part of Classic Peg.

When I shared this silly, joyous image that had stayed with me all those years through that character, I got to see other people laughing about the same things that have made me laugh my whole life. What a feeling that is. It's transcendent.

CLLLAAAAA

AASSIC PEG

Satellite

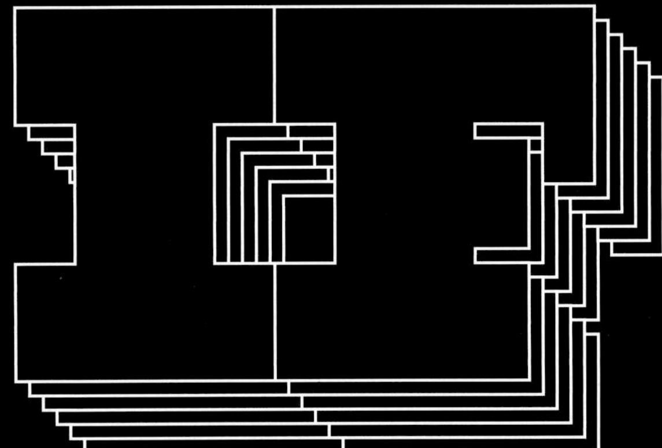

YOU FEEL SOMETHING IN YOUR HEART WHEN YOU LISTEN to my music, laugh at me when I'm being a buffoon, or if you recognize yourself in a performance I gave in a film, it's because whatever that feeling is, I felt it in my heart first, and then I figured out how to share it with you. I don't care if that sounds precious. It's real. It's the realest stuff there is. And we're dealing with make-believe, make-believe that shows us what real is. Wrap your brain around that.

I'm like a satellite. That's it. All I do is receive something. And then my only job is to translate it, to turn a feeling into a physical expression or a sound wave so I can transmit that same feeling.

When you watch a movie, what you're looking for on a subconscious level is a piece of yourself to be reflected at you. When I play a character, I have to find myself in them, too.

It was a milestone for me as an actor when I was cast to play Sean Parker in *The Social Network*, which was written by Aaron Sorkin and directed by David Fincher. I knew that there was something interesting about the way that character was written. He was full of charm and quick wit, but he was saying twice as much with what he wasn't saying.

I had to walk in with respect for the character. All I needed was one thing where I could say, "I have something in common with this character." For Sean Parker, it was easy. Ambition.

That was what touched me.

Whether I'm playing someone who is fictional or I'm playing an interpretation of someone who once lived or is living, I still have to find a real place within them that makes sense to me in order to portray them honestly. Even a fictional character is based on something real. Even an animated film needs that realism.

When we were making *Trolls*, I would go in for recording sessions to voice my character, Branch, for three or four hours at a time. While I was recording, I was being filmed by two or three different cameras to catch all my natural facial expressions as I spoke. This helped the animators make this troll as human as possible, so it can touch you.

There must be a moment when something touches you, in order for a film to work.

If I experience a feeling or an idea or a sound, I trust the way it makes me feel. It's my job to synthesize it and share it in the truest way I can so other people can connect to it along with me.

244
202

244
202

244
202

244
202

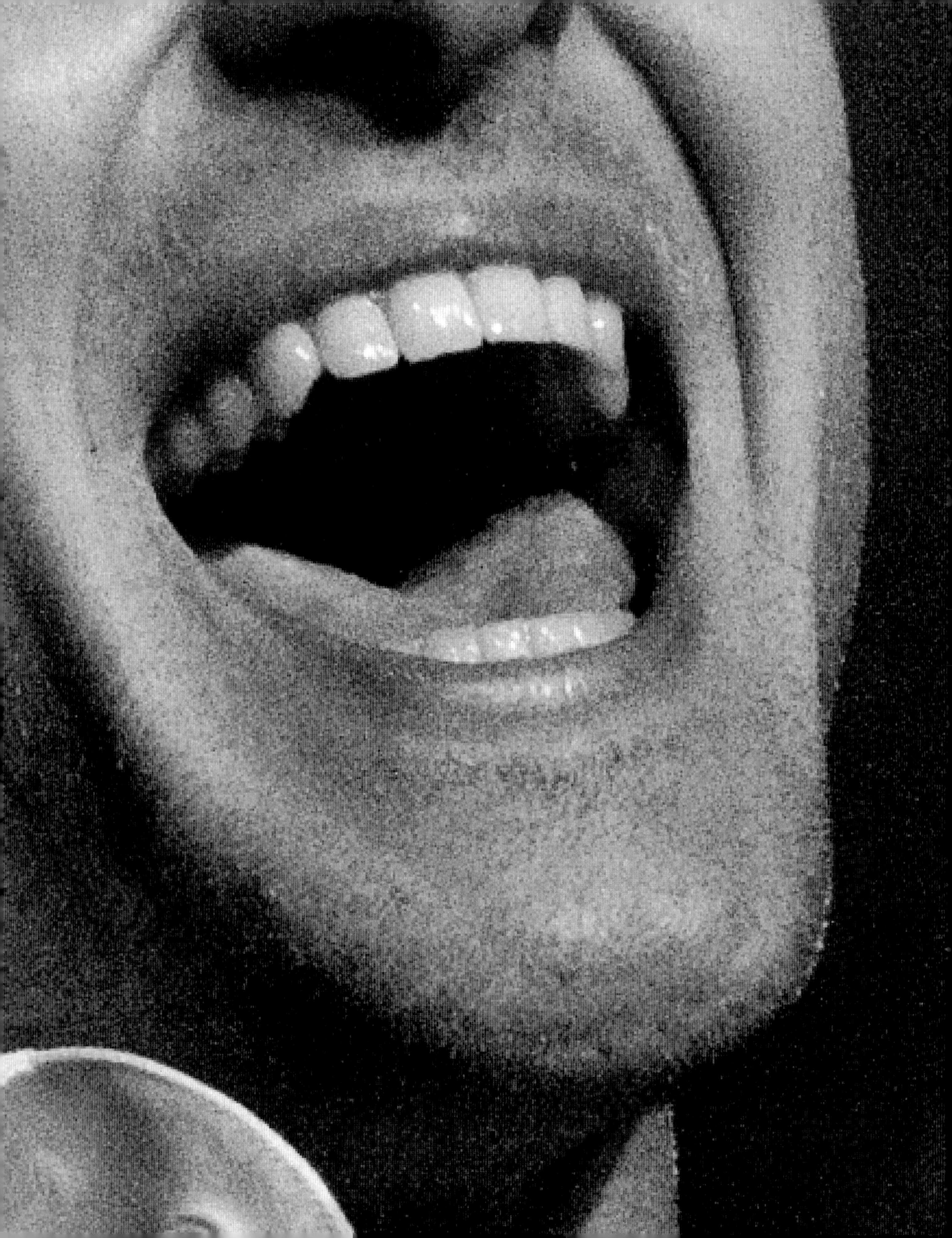

I'm like a satellite. That's it. All I do is receive something. And then my only job is to translate it, to turn a feeling into a physical expression or a sound wave so I can transmit that same feeling.

MAN OF THE WOODS TOUR 2018

MAN OF THE

I'M CEO,

BITCH!

Playing Favorites

HEN I WAS YOUNG, IF YOU had asked me who my favorite singer was, I wouldn't have been able to answer the question. I could tell you what my favorite songs were, for days. But I never wanted to be Michael Jackson. I didn't even know what that meant. When I was little, I heard "Raspberry Beret" for the first time, and turned to my mom and asked her who was singing. But I never wanted to be Prince either. I still don't know what that means.

Maybe that's because Shelby Forest, Tennessee, is a thirty-minute drive to a Walmart. I never went to see Prince when I was a kid. I never saw Michael Jackson in concert; in fact, the first time I saw him perform in person, I was onstage with him at the Video Music Awards in 2001 at the Metropolitan Opera House. I was singing and he came out and just danced during a performance with 'N Sync while I beatboxed. A week later, we performed with him again at Madison Square Garden. That wasn't the kind of show I ever saw as a child; it certainly wasn't the kind of experience I ever imagined. Al Green may have lived in the neighborhood, but as a kid, that whole world, the world of the people making the music, seemed like another planet. What reached into my world, what shook me up, what saved me, was the music they made. The music flooded into my house, and I wanted to be part of all that.

For me, it was all about the songs. It wasn't about who was making them. It was all about the music and how it made me feel. Besides, who can say if Michael McDonald or Michael Jackson was the better singer? They're totally different. That's subjectivity. That's individual preference. That's what I took from my childhood, and it's still influencing my work today.

My job is to make something great. That's it. Some people count achievement by awards; that's not how I think about it. I've won Grammys and Emmys and they don't define me. Just because I have awards doesn't mean I have a trophy case in my house—and I don't; my mom has that stuff. It's nice to be recognized by your peers. Sure, everyone wants to be invited to the party, and it feels that way when you receive any nomination for your work. It means just as much—or more—when I get a letter from someone about a song, or when someone tells me that my music got them through a hard situation, or that my music makes them dance.

I'll tell you this: people don't care about whether a song was number one when they're listening to it in a concert setting. "Mirrors" was never a number one song. I've performed it in front of more than a million people by now, and they know every word, and they're singing it as loud as they can, and it's an experience. It's an experience for them—and for me.

That's why I'm here. That's what I love about my job. I love the spaces that are not about the win but the practice. I think that's why I've started to like golf so much, especially after I discovered that I couldn't be anything else other than a musician, actor, and entertainer. If I'm competing in those arenas, I'm competing with myself.

I don't want to compete, I want to connect. I want to connect to my music and to the audience. I want to do my own best work and support other artists who are doing brilliant work that inspires me.

I was brought up to have a sensibility about what I'm doing and what others are doing. That's what I learned from the way I took in the world around me as a child, when I had the space to hear the music and to listen to songs and respond to the way all of that made me feel. When I think back to how I connected to music then, I know that I wouldn't have cared about Justin Timberlake. But I think I would have liked the songs.

FOR ME, IT WAS ALL ABOUT THE SONGS. IT WASN'T ABOUT WHO WAS MAKING THEM.

JT
JT
&
THE
TENNESSEE
KIDS

I LOVE THE SPACES THAT ARE NOT ABOUT THE WIN BUT THE PRACTICE.

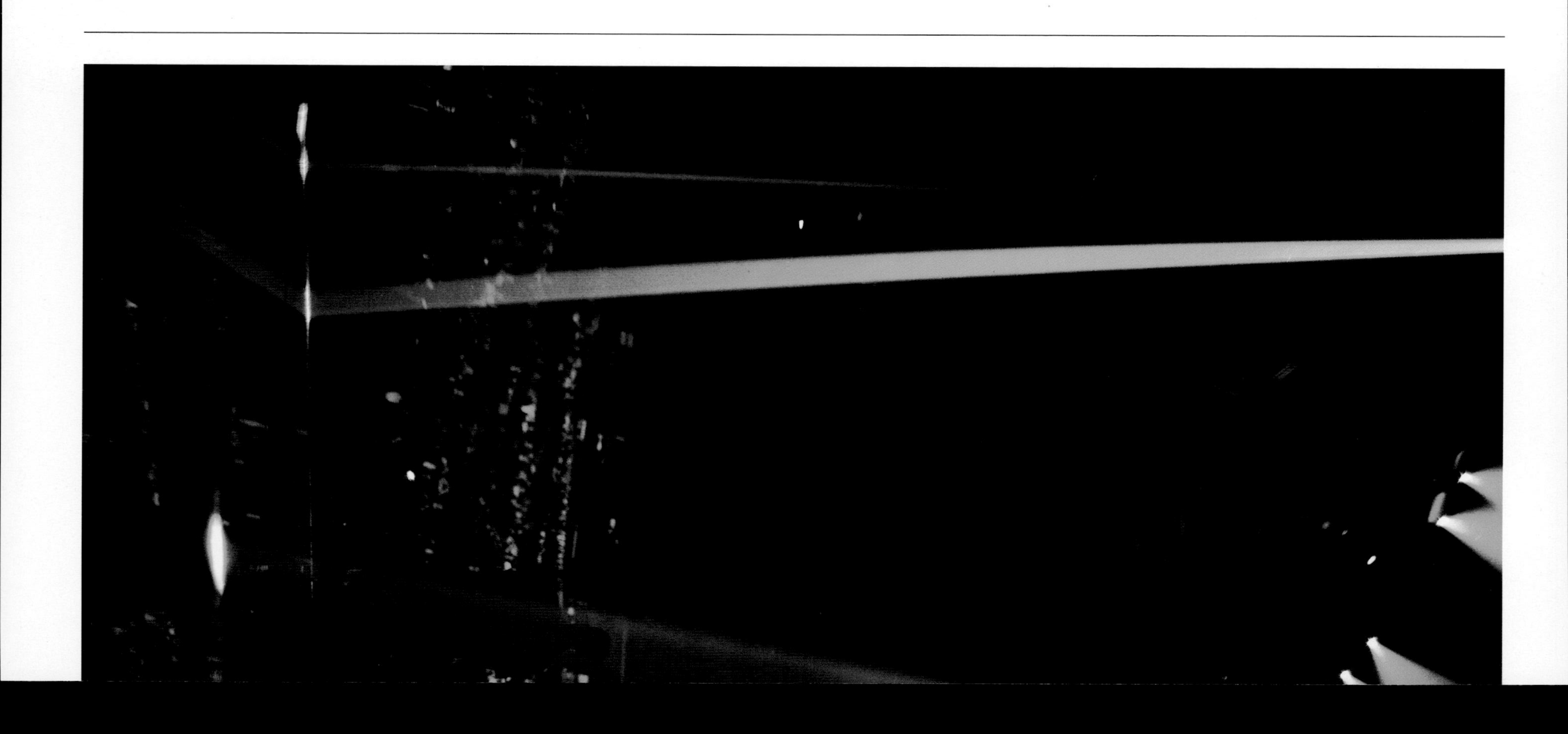

The Art of Performance (Not to Be Confused with Performance Art)

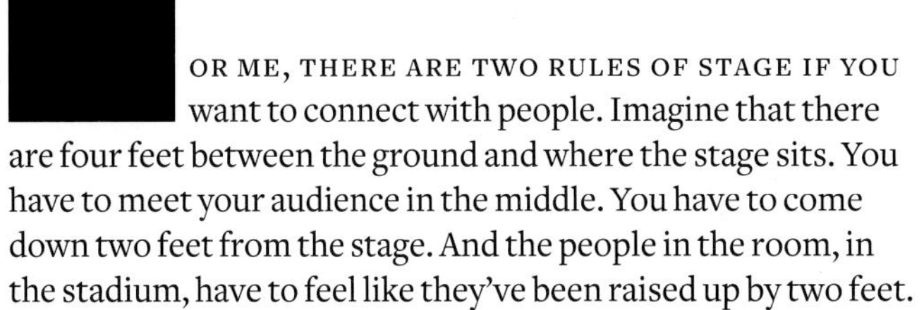

FOR ME, THERE ARE TWO RULES OF STAGE IF YOU want to connect with people. Imagine that there are four feet between the ground and where the stage sits. You have to meet your audience in the middle. You have to come down two feet from the stage. And the people in the room, in the stadium, have to feel like they've been raised up by two feet.

That's what was so amazing about Michael Jackson. He was always good at making people feel elevated. You would watch Michael, and you'd think, "That's magic. I can do anything. I can fly." Which makes sense—Peter Pan was one of his muses. If you watch Paul McCartney, he makes you feel like you are in his living room and he is playing you songs that you are hearing for the first time, and you already know all the words. And you bet your ass you're singing them at the top of your lungs.

You connect with people when you're authentic—when you do things the way that comes naturally.

But you do have to pay attention to where you are when you're playing a live show. When I was sixteen, I saw the Stones play at the Citrus Bowl in Orlando. They didn't play "Satisfaction," and their fans booed them. They wanted to hear the songs they loved. They expected it, and they got most of it. They were still rabid.

Years later I got a call from *the* Mick Jagger to come and play a SARS benefit in Toronto with said band. I was on tour at the time, so it was a bit of a trek to get there, but who says no to Mick Jagger? Not me. When I arrived, I realized that I was the only artist there who would call himself a "pop" act. I had just released my first solo album and there I was, on a bill with the Rolling Stones, AC/DC, and the Guess Who.

Before we walked onstage, I remember saying to my band, "Hey, guys, I can't imagine this is going to go well." And then we prayed.

The following hour of my life was an affirmation that if there is a God, she has a sense of humor. I walked onstage to beer cans and bottles of urine being thrown at me. Bottles. Of. Urine. Apparently, everyone up front and within striking distance had been there for a long time, and they needed bladder relief really badly. And they had been drinking for a while, so there were all those convenient empty bottles to relieve themselves into.

My opening song was "Cry Me a River," and we made it through that song without me or anyone in my band being hit in the face with anyone else's disposal. I'll call that a win.

Here's what I remember most from that moment. My best friend, Trace, was there, standing side-stage. I glanced over and he looked sort of sick, like his heart had dropped to the floor. It was almost as if I saw him wishing that he could be up there instead of me; that's when I decided that no matter what, I was going to finish my set. Also, this was a benefit festival for a communicable disease. Who throws bottles of urine?

Later, when I was sitting with Mick, he told me that the people responsible were "those silly AC/DC fans." I didn't believe him, until I went onstage with the Stones to perform "Miss You," and the audience threw bottles again at Mick and me. Here's something you should know. When Keith (*the* Keith Richards) saw what was happening, he walked to the front of the stage, banged on his chest like an angry gorilla, and challenged the culprits to throw something at him. Said culprits cowered and decided it was not in their best interests to continue.

This is all true.

Note to self: an arena can create a mob, and you can't control a mob. Once you lose a mob, you're playing with fire at that point. Good luck—and duck.

As a performer, I always want to understand where I am performing and who I am performing for, because I enjoy giving people what they want. When it comes to a live show, I don't mind being a servant in that way. Getting up and playing to entertain yourself is not connecting with the people. Do that at sound check. People did pay good money to come see you, and they didn't come to see you read *The Great Gatsby*.

I'm not Andy Kaufman, who did that once. In one of his stand-ups, he began to read *The Great Gatsby* out loud. The audience was not pleased. But that was his mission. That's what he wanted to have happen. Man, he was brilliant. But I'm not here to make people leave. I'm not here to get bottles of piss thrown at me. I'm here to perform. I'm here to connect.

I feel lucky to have found something that I can do that links me to people all over the world. It's a crazy thing. I write a song in a room and I record it and I sing it the way that I sing, because that's the only way I can really do it. Twelve months later, I'm in an arena and I'm performing. I can never forget that I was there for the creation of this song when I look out and see all these people from all walks of life, and the whole room is singing. All these people, there together, singing a song that I wrote.

It's a heavy thing, man. It's enough to get really high off of. I've even been moved to tears. That was another life-changing moment for me: when I realized that people didn't only listen to you sing, that they would sing with you, too.

JT

TL

TO CONNECT WITH AN AUDIENCE, YOU HAVE TO COME DOWN FROM THE STAGE.

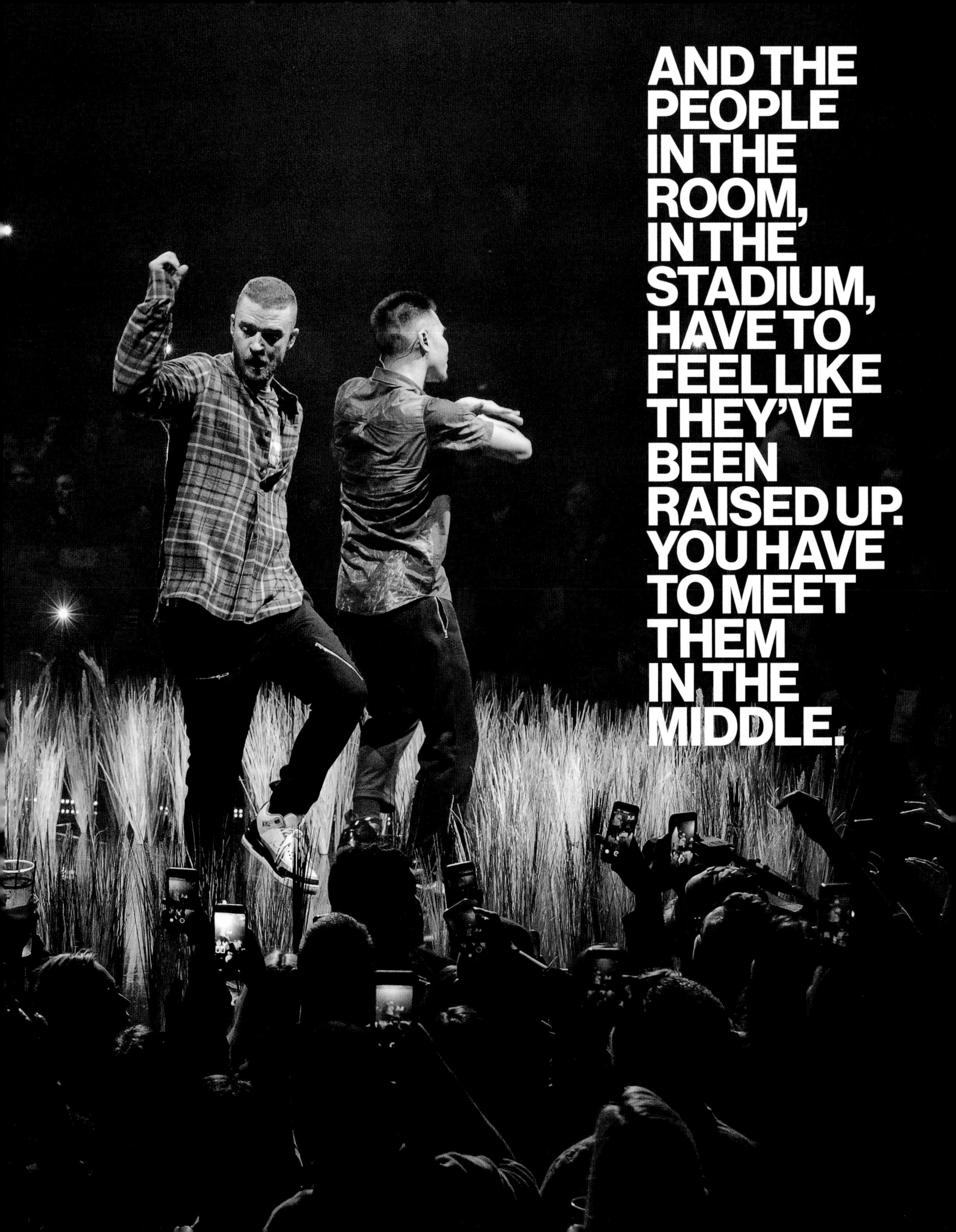

AND THE PEOPLE IN THE ROOM, IN THE STADIUM, HAVE TO FEEL LIKE THEY'VE BEEN RAISED UP. YOU HAVE TO MEET THEM IN THE MIDDLE.

TOUR

SERVANT

WENTZ

The Practice

“I don’t know where I’m going from here, but I promise it won’t be boring.”

David Bowie,
at his fiftieth birthday concert
at Madison Square Garden,
New York City, 1997

I'm constantly trying to understand who I am so that I can turn the inspiration in my heart into something tangible, into a feeling that can be experienced by someone else. Sometimes, it ain't easy. If you want to make something good, you have to work for it. You have to work with people who know how to work hard, and you have to work harder than everyone else. You have to be open to being inspired by other artists and your own experiences. You have to give it everything you have, and then give a little more. You can't make excuses. You can only make the time. I have been in rooms full of people staying up all night and working a song, working a bit, working a scene, until it feels right. I have seen how an idea grows and changes and takes form over and over again, as many times as it takes to turn it into what it is longing to be, until it becomes the thing you want to see or hear or laugh with. If you want to do something, if you want to make something, if you want to create something, I've learned that you can't be afraid to do it wrong. You have to dare to suck. It's not about right or wrong. It's about what's most interesting. Every movie that makes you cry or laugh, every song that makes you dance or feel romantic or nostalgic, every comedy bit you watch again and again, and then repeat the punch lines from to your friends, if any of these things make you feel amazing, if they remind you who you are, remember: it takes a lot of hard work to make it look easy.

ALL THAT
ALL THAT
ALL
ALL THAT
MATTERS
MATTERS
ALL
THAT
M ATTERS
IS
I S THE
SONG
SONG
SONG

All the Beautiful Accidents

HEN I WAS SIXTEEN, AND singing and performing with 'N Sync, our record label sent us to Stockholm to work with the producer Max Martin, who had all these hits with Ace of Base, Robyn, and the Backstreet Boys. Going to Sweden was an exercise for us to record with this huge European hit-maker to jump-start our career as a group.

Working with Max was an education in music. Now I was learning how both songs and vocals were produced. Singing in the recording studio with Max, with a microphone two inches from my face, I had to sing in an entirely different way to interpret the songs. It wasn't at all like being onstage, working to project myself out to the audience in the last row.

Having that experience at such a young age, where I was so impressionable, really made an impact on me. Max is influential and he's humble, and he cares about the right things, by which I mean, the song. All that matters is the song. That's a philosophy that I've really adopted in my songwriting and my producing. When I saw what Max was doing, I wanted to be part of it. Watching him work made me want to do what he was doing. That was when I really fell in love with songwriting. It just felt genuine to me, to help craft the song. It still feels that way.

I started playing instruments around that time because I wanted to learn how to write songs. I took a few guitar lessons when I was ten because I wanted to play music, but I quickly realized that I didn't have the patience to learn how to play just to play. I was not a patient kid. It wasn't until I really fell in love with songwriting that I delved deeper into playing instruments, because I wanted to have a deeper understanding, from a production side, of why certain guitars sounded the way they did. I wanted to understand all the effects of the different instruments on a song.

When I'm writing songs, I want to be looking at the forest, not swinging from the trees. I find that I have a better relationship with what the arrangement should be if I'm sitting and listening to someone while they're playing what I wrote. It gives me perspective. I'm thinking about how the pattern of the guitar relates to what I'm writing or the groove, instead of sitting there trying to figure it out. My brain can do something else—it can add to the part, or change it. Also, a different human playing the same part is going to naturally sound different. If I write a rhythm line on a guitar and then have my guitarist play it, it's going to have a specific cadence, timing, and feeling. That's where the beautiful accidents, as I like to call them, can happen. Maybe the guitarist heard the part differently when I was explaining it or singing it out loud, but then plays it in a totally different way, and it unlocks a whole new melody. Beautiful accident.

I'm not there to riff on the guitar. What I'm looking for is how the instrument works into the song. The instrument becomes a proxy for the song. The song is what means something to me—and how I can make it mean something to you.

RIVERA

RAMLÖS

WHEN I'M WRITING SONGS, I WANT TO BE LOO KING AT THE FOREST, NOT SWING ING FROM THE T REES.

Write by Heart

I DON'T WRITE MY LYRICS DOWN.

EXCEPT, SOMETIMES I DO. THERE AREN'T ANY RULES. THERE are so many ways to get there.

When I'm writing a song, most often the process is all vocal. I pace around the room, collecting words that make sense to me, letting how I feel color the words I choose. That's how I always write. It isn't a technique I learned—it's instinctive.

Every moment of creation starts with inspiration, whether its roots are conscious or not. When I'm writing in a stream of consciousness, letting my mind wander, I'm inspired by something.

I've had different experiences while writing. There have been times that I've written a song, and as I was writing it, I was conscious of why I had been inspired to write it. Other times, I've just written a song and only later understood what I did and why and where it came from.

Everyone has unconscious sources of inspiration. They're always there, feeding you, whether they come from the world around you, from the work of other artists, or from childhood. You might not know exactly how you metabolize those things, but you do, and you do it reflexively, like blinking or breathing. But when you pull from that pool of images, feelings, and experiences, channeling those references into an art form, you begin working with what's inspired you on a conscious level.

One thing I've learned is that I have to let the feelings or ideas that inspire me become what they want to be. I have to follow them. I have to give up my ego. I have to be humble. I have to be open. I receive all that stimulus, taking a feeling that is personal to my experience, and even if I might not be able to name it, I translate it into a song.

Then I just get out of the way.

I've been scorned. I've been pissed off.

That feeling inspired me to write "Cry Me a River."

I wrote "Cry Me a River" in two hours. I didn't plan on writing it. The feelings I had were so strong that I had to write it, and I translated my feelings into a form where people could listen and, hopefully, relate to it. People heard me and they understood it because we've all been there.

But it's more than that. For a song to work, it has to register, which is why the kind of songs I write use repetition and a hook. They call it a hook because it hooks you, but really, the verse should be a hook. The chorus should be a hook. And I think the melody should hit you in a way that makes you remember it.

A song is three minutes long. If you're a heavyweight boxer, three minutes is an eternity. If you're a songwriter, three minutes goes by in a heartbeat, and you only get two shots: that first listen that gets the song into listeners' heads, and the second, which calls up what they heard the first time and, hopefully, makes them remember it. If it happens the third or fourth time, that's the icing on the cake—because now people recognize the music, they get to participate. They get to sing along. They get to make it their own.

Songwriting can't be rushed, but there are moments where the rush comes in. You have to be a satellite, able to receive what's coming in. And then you have to respect what you're hearing enough to go, "Wait. Is this it? Or was this leading into something else?" Sometimes what makes a song recognizable is paring it back to its elements. That's why arrangement is so important. You have to understand what makes it resonate.

Sometimes it's about adding something unexpected that ends up being the most vivid part. If you listen to "Eleanor Rigby," in my opinion, it's important to remember that Paul McCartney originally wrote that song on a piano. In the early version, her name wasn't Eleanor Rigby either. It was Miss Daisy Hawkins. But in the final version, what you hear is the opening accompaniment of staccato strings with the opening line, "Ah, look at all the lonely people," which is immediately recognizable and part of a melody that stays with you. The strings are so important to that song—and they're so good. They're like a guitar riff in a rock song: sometimes that one melody is all you need.

Unarranging is part of arranging.

That's all part of the refining process. I have to be willing to go with the song on its journey. I'm just along for the ride. I'm not dictating it. I'm just there. So when that one part calls out, I'm ready to hear what it wants to tell me.

When I record a song I write, I literally go for a ride: I listen to the demo in the car. That gives me the space to listen to the song, to think of a way to adjust it, to try this or that. The puzzle is right there. I have to be able to identify the pieces that can connect each other, until it gets to the point where I feel like I can let it go.

As I've grown, and my abilities and confidence have gotten stronger, I've gotten to the point to where I can let go of a song more easily. I'll beat it up before it even gets to everybody. That is my job. That is where I have to be conscious. Writing a song isn't something you do to please everyone. You have to do it because you feel that you are on the path to something great.

It's a practice. It's not a goal.

When I've turned over every stone, even if I end up back at the place I started, when I know that I've tried every sort of pathway with the sonic of this song, and this is what it is, and it transmits the feeling I wanted it to transmit, I'm finished.

Songs are not finished when they are perfect. They are finished when you feel like they are done.

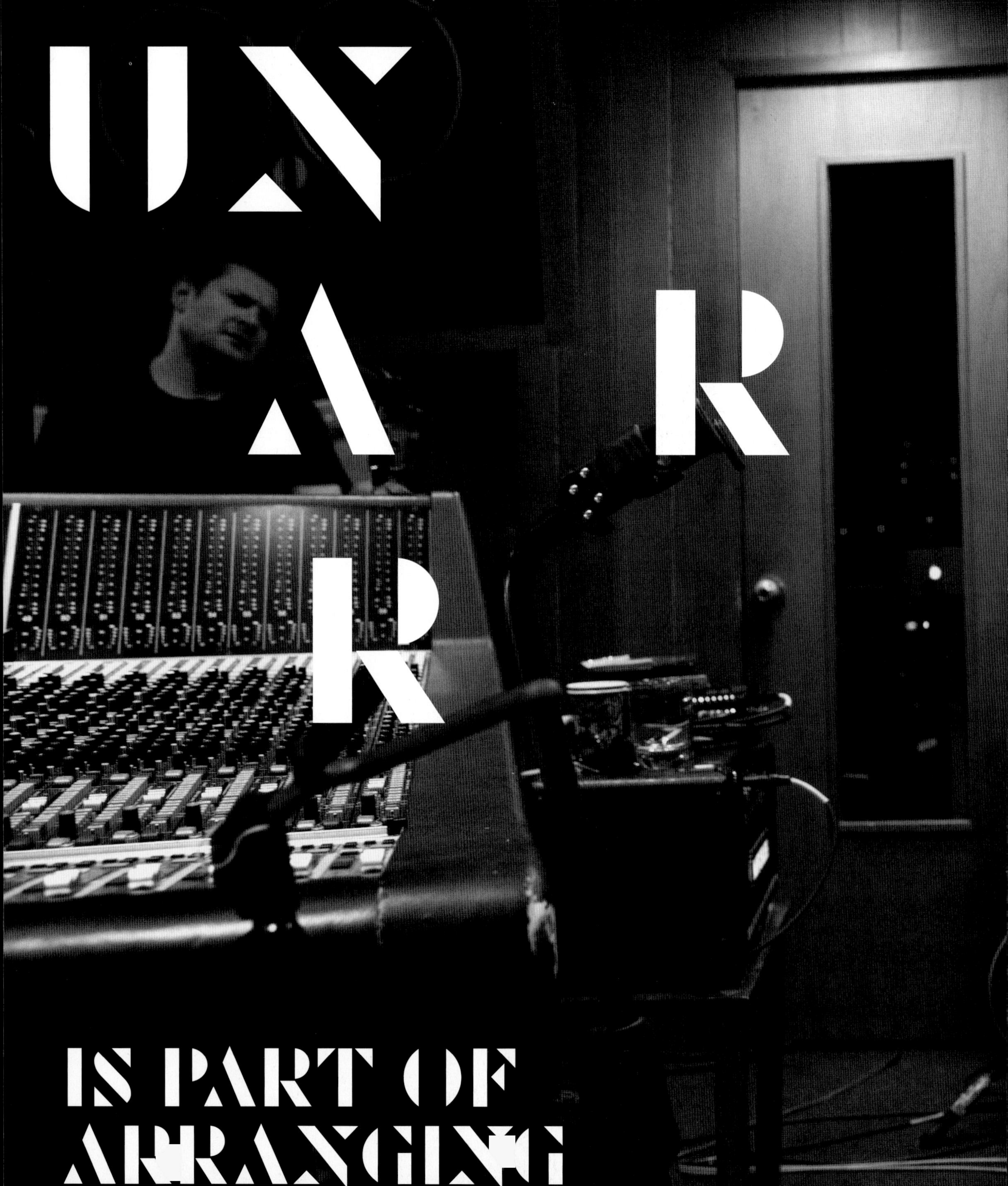
UN
A
R
R
IS PART OF
ARRANGING

A
N
G
I
NG
FIRE
NATIVE INSTRUMENTS

THE STORY OF HOW I WROTE "SEXYBACK" HAS BEEN told a million times. Maybe you haven't heard it. Here goes: Tim—Timbaland. Yeah. You know who—made the beat. It was sparse, super sparse. I fell in love immediately. I said to him, "Oh, I know how I'm gonna come in on this song."

And then I just blurted out the line.

"I'm bringing sexy back."

And Tim went, "Yeah!"

I should add that it's impossible for Tim to do anything out of rhythm. The man could trip while he was walking and it would still be in rhythm. His exclamation became what you might know on the song as what seems to be everyone's favorite part to scream out in a club when they're completely inebriated . . .

"YEAH!"

That's the story that's been told.

But that moment happened because I had been listening to David Bowie for two weeks straight beforehand. I was listening to "Rebel Rebel," which is about cross-dressing and going out and being completely comfortable, being a man or a woman, no matter if you are a man or a woman. At least that's what I took from it. And when I read the words, I saw clearly that the unapologetic lyrics and the vibrant sound that goes with the content are completely synchronized. They're feeding each other.

That's what I wanted for "SexyBack."

This is an example of the times that I'm writing a song and I know where it's coming from. When I wrote "SexyBack," I knew right away where it came from, and I was very clear on what I wanted it to be. I wanted to create something that made me feel like David Bowie made me feel when he sang, "Hot tramp, I love you so!" (With the exclamation point.)

That to me is a great example of conscious inspiration. The audience is just the watcher. I'm just the listener. I'm just the receiver. What is inside comes from outside, but it is tempered and changed by that process of awareness as I experience it and translate it back.

With "SexyBack," I felt like I was making my version of rock and roll. Tim and I were clearly making dance music—our version of it. You can't write a song like "SexyBack" and have violins on it.

I was looking for something that felt wrong, that felt raw. I was looking for something with an attitude toward it, within it.

I grew up with a ton of different influences. I got to experience the heart and spirit of rock and roll. But I also got to listen to disco, new age, and hip-hop. All that unconscious creativity sits in your bones for years, and for some reason I could identify all of them at the same time and weave them together. My unconscious fed me as I made conscious choices about how to make "SexyBack" have the right feeling.

It didn't hurt that I was working with Timbaland, who is so famously great at sonic. The man knows what to do with a sound wave: he's the best I've ever seen at layering drums. It's what he does. Since I was a kid, sounds like that have always sounded better and more exciting to me.

Tim knows when to make noise and when to leave space. That was the other thing about "SexyBack." I wanted to keep it spacious. I didn't want to fill it up with a bunch of whatever. The thing that makes music interesting to me is the space between each note. Like Miles Davis said, "It's not the notes you play, it's the notes you don't play." It's the space between the notes that makes the music. When I first heard that quote, I felt that I should aspire to continue to make the sound progression interesting, so that the puzzle pieces can catch us. That's how rhythm works. Like pieces of a puzzle. We are always trying to make sounds fit together. If I were to tap out a rhythm, even if you have no rhythm, your brain is going to respond to it and process it subconsciously. If I were to change the pace and add space, more space—

—now you're really listening.

That's because space keeps the neurons continually interested. You have to be patient. Space makes you wait for what's coming next. That's why I love to sit on a beach and watch waves. Because every wave, no matter what rhythm they're all coming in, is going to be different. And we wait for each one. "SexyBack" had sound and it had space, and when we played it in the studio, we loved it.

What makes somebody love a song?

What makes someone hit replay?

What makes somebody dance?

What makes a song work or not?

When I was twenty years old, I asked Stevie Wonder about *Songs in the Key of Life* and *Talking Book*, and how he continues to be inspired. He said, "I don't know, man. There are two kinds of music: good music and bad music. I try to stay away from the latter."

So "SexyBack" sounded good to me. It sounded good to Tim. And the proof showed itself every time we put the record on.

Only the record company didn't feel that way. They said, "This is never going work. It doesn't sound like you."

But music is subjective. I can't walk into creating anything by thinking that I'm there to make something that'll please everybody. That's the fastest way to make something boring. I can't think about if everybody's going to love it. I just need to love it.

"SexyBack" could have gone another way, but I knew it wasn't going to. That's a learned technique, to understand your form. My instincts told me it was going to go the way it went. For every reason I was given why it wouldn't work, I knew it would.

Every obstacle is an opportunity, which is why I don't think there are any bad ideas. You can't break through without conflict. My mom once admitted to me that when I was younger, she'd make sure to tell my teachers the secret of getting me to do something: tell me that I can't do it.

I was always up for a challenge then, and I'm always up for a challenge now.

The record company challenged me.

They said, "It's too fast. You're not even singing."

But that was what I wanted. What we made was exactly what I wanted it to be.

I said, "This is why it's going to work. It doesn't sound like anybody. But it sounds like somebody."

What I meant was, I wanted something on the radio that just sounded like nothing I'd ever heard. I was excited about that. I put that effect on my voice for a reason. I wanted my voice to sound distorted. I didn't want people to know it was me, because I wanted people to say those words, and feel like they were whoever that character was in that song, however they imagine him to be. If they imagine him as Ziggy Stardust or James Brown. I didn't want Justin Timberlake to be a part of this song. Any association with me is not what this song is for. I didn't want it to be me singing. I wanted it to be you.

I wanted to have so much force, I wanted you to connect with me so hard that you take off all your cool, and you just live. That's what "SexyBack" does. That's why it worked. It made everybody take off their cool.

I really believe that "SexyBack" became what it became because it was so different from everything that was on the radio at the time. It was like a coloring book. The lines were there, but you got to fill in the colors yourself. You got to make up your own idea of what it was supposed to be.

That's why it was so important that I didn't sound like myself.

That's why the proclamation is so important as well.

Because when we were done, when you listened, when you "brought sexy back," you really brought it.

YE
SCOTCH
AND
SODA

AH

The City That Never Sleeps Never Lies

I LOVE NEW YORK CITY. It truly is the greatest city in the world—in my opinion. One of my favorite things about this place is that New Yorkers can't lie. They're so honest they might offend you with the truth.

I can always tell the songs or *SNL* sketches or films that I've been a part of that register with the masses more than others when I'm walking the streets of downtown NYC. I'm talking about straight-up catcalling.

Them to me: "Hey, JT! Loved you in the Facebook movie."
Me to me: It's called *The Social Network*. Whatevs.
Me to them: "Thanks, man!"

Them to me: "Hey, JT! I got my dick in a box."
Me to me: ???
Me to them: "Please don't put that there. It was just a joke."

And then there was the time when I crossed paths with a UPS delivery guy who shouted out, "Hey, JT! I'm bringing sexy back!" This was particularly funny to me because we all know what a UPS uniform looks like: his shorts . . . were . . . short. Really short.

If he truly felt that way, at that moment, in that outfit, then damn it my instinct was right on that song, and I was creating something that wasn't just a proclamation for me but for anyone and everyone.

If hindsight really is 20/20, at least the UPS guy wasn't claiming to have his dick in a box. That would just be scary.

Writing Rooms Are the Rooms You Write In

ONGS ARE EVERYWHERE, JUST LIKE SOUNDS ARE EVERYWHERE. Sometimes when I'm sitting in a restaurant I like to stop and just look around the room. I wonder to myself, *That couple in the corner, what's their story? They're having white wine at 4:00 P.M.; there must be something to celebrate. Or something to hide. What about that dad and his daughter on the other end of the room? Oh, they just stopped in to get some chicken nuggets. But where were they before that? Did he take her to the playground before they stopped in for a snack?*

I write songs wherever I am, whenever the inspiration strikes. I've written lyrics while watching people. I've written lyrics sitting alone in a field, watching the sun sink over the trees. I've come up with time measures from hearing rhythms in the world while I walk down the street.

Inspiration comes when it comes, and I have to be open to it.

I was checking into a hotel in Philadelphia when I missed a call from will.i.am from the Black Eyed Peas. We had known each other for so long, and we had talked about writing something together and making a record that would be a positive thing in the world. I listened to the voice mail as I walked in, setting my bags down. The voice mail was tagged with an instrumental of a staccato string line on top of a mid-tempo beat.

If you've ever tried to play someone music over the phone you probably know that there's only certain frequencies of your sound that the speaker on your phone can transmit. So the kick and the snare were very faint, but I could hear them enough to get a sense of where the downbeat started. What was really coming through the speakers was the progression of staccato strings that were laying out a melody in my head.

I immediately restarted it and played it over and over again, maybe fifteen times. I listened on speaker setting and back to headphone setting to see if I could get other frequencies of sound to transmit.

On the flight to Philly, I had been listening to Marvin Gaye, and as I listened to Will's track, a breezy falsetto melody started playing in my head. That's why I always say that songs are a gift—because ten minutes later I had the fully realized chorus of what would go on to be known as "Where Is the Love?"

SONGS
ARE
EVERY-
WHERE.

WILL
SING
FOR
FOOD!

You

have

to

dare to suck

WROTE THE SONG "MIRRORS" for my wife. We were living together at the time. We weren't engaged yet. In fact, it was years before I proposed.

The song didn't come out until just before we got married, years later. The video became a dedication to my grandparents. I learned about long-lasting love from them.

I know that's what I have with my wife.

When I met Jess, it was undeniable how beautiful she is. We were at a surprise party in Hollywood at a speakeasy kind of a bar, a private club, and we were standing around in a group of people. I made some sort of sarcastic comment, really dry. Nobody got it except her. She laughed, and I noticed, all of a sudden, and in the way where you wonder if a person's like you, if they have a very dry, dark sense of humor, too.

We talked that night. The DJ played "Lucky Star," and we danced. And then she was gone.

I didn't ask for her number. It wasn't the time. But I was thinking about her. Thoughts kept leaping into my mind: *I'm kind of really interested in that girl. There was just something about that girl. Kind of interested in that girl. Goddamn it.*

I had to psych myself up. I had to remind myself that I was me—giving myself a pep talk, shadowboxing with myself, rubbing my own shoulders.

Time went by.

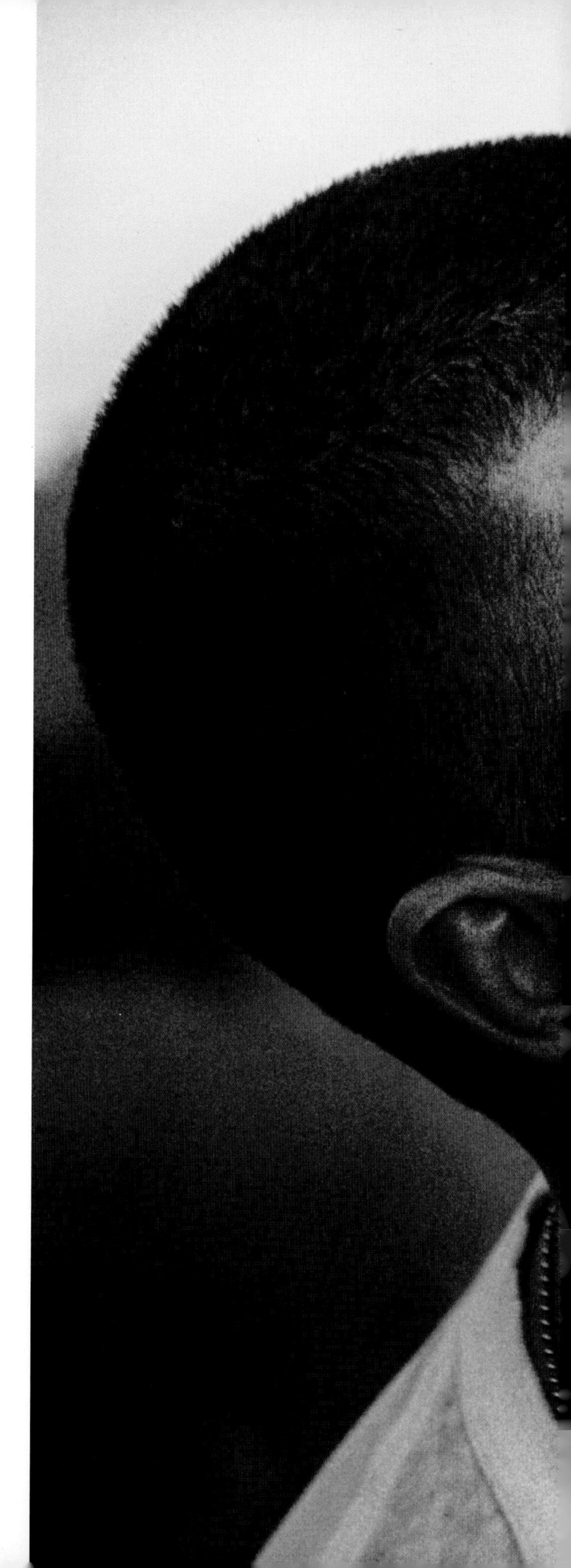

The Other
Half of Me

My first show for the *FutureSex/LoveSounds* tour was in San Diego. A friend who also knew Jess was going to come to the show, and she asked me, "What if I was to bring a certain somebody?"

I said, "I could be into that." Patience may not be my strong suit but, apparently, playing coy is.

My friend brought her to the show. They were with a whole crew of girls, and they all came into the dressing room and hung out. After the show, they were going to go back to LA, and I was going on to Anaheim.

I asked them if they wanted to come with me.

"I'll give you a ride, you can come on my bus. If you want . . . "

And they did. Jess and I talked the whole way up, joking around. Before she got off the tour bus, I said, "Can I have your number?"

That was when we started talking on the phone. We didn't date at first. For about two weeks, we talked on the phone, because I was on tour.

Then the Golden Globes came up, and that's when we planned on seeing each other. From that moment on, we started dating. We were both still seeing other people, keeping ourselves safe from getting hurt, from really putting ourselves out there. It took a bit for both of us to admit to ourselves that we were really, really that into each other.

When I came back from tour, we spent a month together. After that, I said, "I really want to be exclusive." And, somehow, she said, "So do I."

It was crazy, because I was about to leave for Europe for three months. Two weeks later, I came back to LA to see her. Then she flew back to Europe with me, and we spent two weeks on the road together. We were going to different cities, having a ball, with room service and pay-per-view and great dinners. And Scrabble.

We've had a lot of unforgettable times. She's become a huge influence on my life, and I have such admiration for her, especially seeing her as a mother now. But I had admiration for her before. She's a very good writer. She's a poet. She's a tremendous actor. She's funny. Very funny. And she's one of the most patient people I've ever met. So, could finding someone who covers for your shortcomings be a thing that plays well for relationships?

She changed me. She changed my life.

All of that is in "Mirrors." Not the details. But the way it felt to have my life be touched by her.

Al Green, This Time in a Bubble Bath

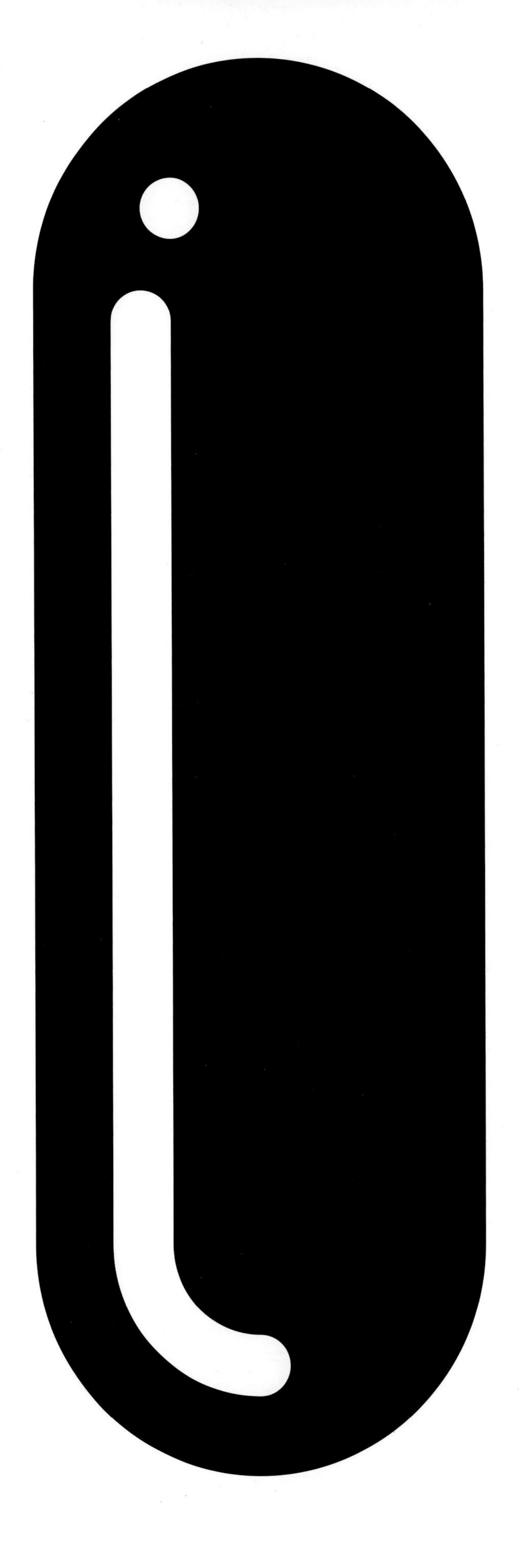

WAS AT THE GRAMMYS TO PERFORM, IN 2009. WHEN I ARRIVED at my dressing room, everybody who runs the show was in there.

I said, "Uh-oh, what did I do?"

It wasn't about me.

One of the acts wasn't going to make the show. They asked me to come up with something. Anything. At the last minute. For a live show.

I said, "Get me a list of presenters."

I saw Al Green on the list. Boyz II Men. Keith Urban.

It was lunchtime, and the show was going live that afternoon, so I only had a few hours to create a performance. I thought who better than Boyz II Men to come in and put their own spin on a vocal arrangement. I asked Keith Urban to play a guitar solo. We called Al Green, but we didn't know if he was going to make it.

We rehearsed a couple of times—someone kept running in with updates about Al Green.

"We called him."

"We're trying to reach him."

"Al Green's people have called us back."

"We think he's a half hour away."

And then—

"We think he's a half hour away."

Twenty minutes later—

"We think he's twenty minutes away."

And then Al Green comes running up to the stage at Staples Center, wearing a mink coat. He said, "I'm so sorry I was late, Justin. They called me, I was in the tub."

I got this mental picture of one of my heroes, in a bubble bath, and he gets the phone call. Runs into the closet, bubbles still on him. Suddenly, he's out of the closet, fully dry, fully dressed, with his white suit and his mink coat, like his Superman cape.

I said, "How about you do this verse and I'll do this verse, and we'll sing this part together, and then you . . . you know, just like we did in Memphis at the New Daisy."

This is where you have to be able to trust yourself. You have to be able to trust yourself enough that you don't have to rehearse that much. You have to have that stillness within you, to be able to trust all the work that you've done before, all the practice that you've done before, so you can just get up there and know that whatever happens, it's going to be good. Or it's going to be interesting, at least.

We did it, and it went well. Really well. Like, standing-ovation well. Note to all entertainers: if you want a shot at a standing ovation, call Al Green.

Don't Be Cool

Do you have that one friend who, whenever you get together, you make asses of yourself in public, you create jokes that only the two of you get, and that other people possibly find annoying? You probably encourage each other to drink too much, thereby encouraging the creation of more jokes that nobody else will get and probably find annoying, too? I have a friend like that. His name is Jimmy Fallon. I guess the only difference between my annoying friend and yours is that mine has a TV show where we can live out this idiocy in public for your comedic pleasure. Every collaboration has its own chemical formula. Each person brings what they have to it, and then you see what kind of charge you give each other. If the reaction is greater with both of you there than it would be otherwise, you're onto something. I think that's why people love to watch Jimmy and me together. That's at the root of why I think people enjoy our bits—because you feel like you're watching two friends. And you are. It's true that we think a lot of the same things are funny. But more than that, we find each other endlessly amusing. That's part of our chemistry; we're both completely in. Nobody's trying to be cool. That's what I like, people who aren't cool. Don't be cool. Be angry, be upset, be passionate, be happy, be amused. Cool isn't a thing. It's an unthing. It doesn't exist. The coolest people are the ones who are doing something they're excited about. Jimmy and I both carc. Wc'rc in, 100 percent, 200 percent, trying to keep the bit going, trying to make it funnier, whether there's a camera in the room or it's just us. We're nonstop when we're together. We just take turns being fire and kerosene. We can't stop. Jess says it's the fastest game of Ping-Pong she's ever seen, almost impossible to keep up with. When I'm hanging out with Jimmy, we're doing bits even when we're alone. We just can't help ourselves. It's our natural impulse when we're together. And if I can make Fallon laugh when it's just the two of us, that really makes me feel great. As a matter of fact, most of the bits you've seen us do on television come from us doing that same bit when we're just hanging out. I remember when Amy Poehler was making her return to *SNL* as a host and I was in New York. She had asked Jimmy to make a cameo appearance and then called me up to make a cameo as well. After *SNL* was over, and maybe a couple of gin and tonics later, Jimmy started mentioning "that one kid on YouTube who had performed the whole history of dance." That led us to a seriously unserious conversation about why there was no history of rap. Which led to a rap competition. Which became Jimmy and me rapping classic hip-hop songs from the seventies and on. Which became "The History of Rap." We also have some bits inspired by a bottle of Jägermeister. Don't worry, you won't be seeing those on TV.

BROS BEING BROS

BATTLESHIP
THE ORIGINAL FIGHTING ROBOTS!

COOL ISN'T A THING.
COOL ISN'T A THING.
COOL ISN'T A THING.
COOL ISN'T A THING.
COOL ISN'T A THING.

IT'S AN UNTHING.
IT'S AN UNTHING.
IT'S AN UNTHING.
IT'S AN UNTHING.
IT'S AN UNTHING.

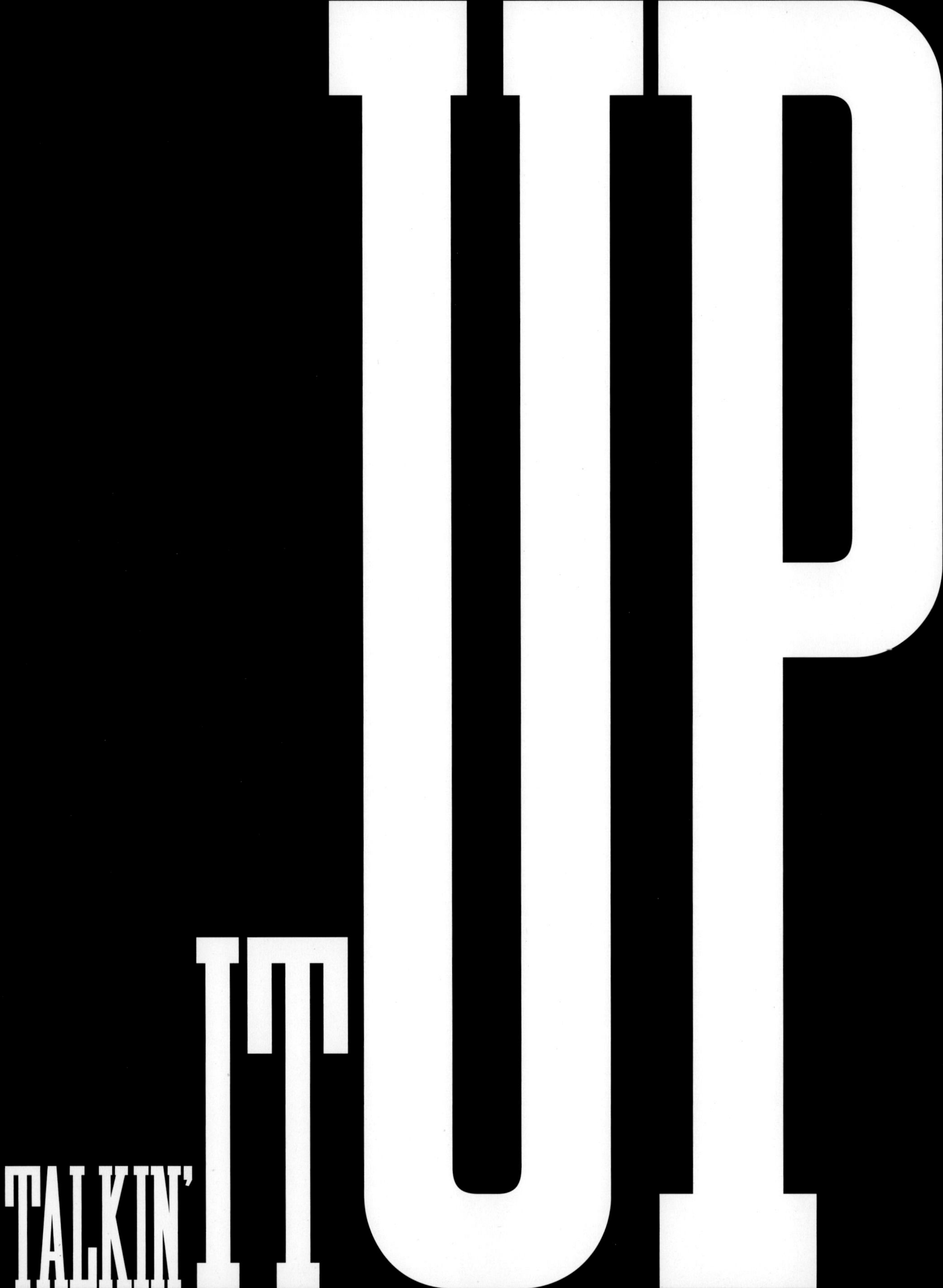
TALKIN' IT UP

HASHTAG

LOLOLOLOLOLOLOL

I woke up one morning and called Tim. I hadn't spoken to him in a while.

He said, "Hey, what's up?"

I said, "I'm ready."

He said, "Ready for what?"

I said, "You know what."

THAT WAS THE BEGINNING OF *THE 20/20 EXPERIENCE*. I put together a group of other producers and musicians, and we congregated at Larrabee Studios in the Valley and wrote for a month. I had two rooms going at the same time. We would go write a verse in one room and then leave it. Then we'd go into the other room and write something else. And then we'd go back to the first room and see what was there, to see what we could connect to the other thoughts that were flowing. All these thoughts, these rooms—it was like writing camp. It was a cool way to write. That's where all the songs came from.

Except for "Mirrors." I had held on to "Mirrors" for years, waiting for the right time.

Now it was time.

Songs used to be short. They had to be short because when you had records made on vinyl, there was only so much physical space. Same for cassettes.

The albums people were recording when I was a kid were limited by the recording technology of the era. Today, we are digital and we have infinite space. Now songs and albums can be as expansive as you want them to be. And that's what *The 20/20 Experience* became: a nonlinear exploration of sound, feeling, and colors. I kept writing and recording, making the album more and more expansive. I liked how long it was. I liked that it just kept going on, that parts that came up in the arrangement could repeat, but then I could introduce a new part that could play off of the first. You need space to do that.

To work for the radio, songs need to be a certain length, but I didn't want to cut the songs down when we were producing them. I thought we'd figure it out when it was time to figure it out. I didn't want that to influence the way we created the songs.

Take "Mirrors." "Mirrors" has a whole outro. When it was time to edit it for radio, there was the thought that we should cut it down so it would be one composition, shorter, including the outro. But we didn't. "Mirrors" was the song. The outro was another song attached to it. They're symbiotic.

If you want to hear both parts, you can listen to the album.

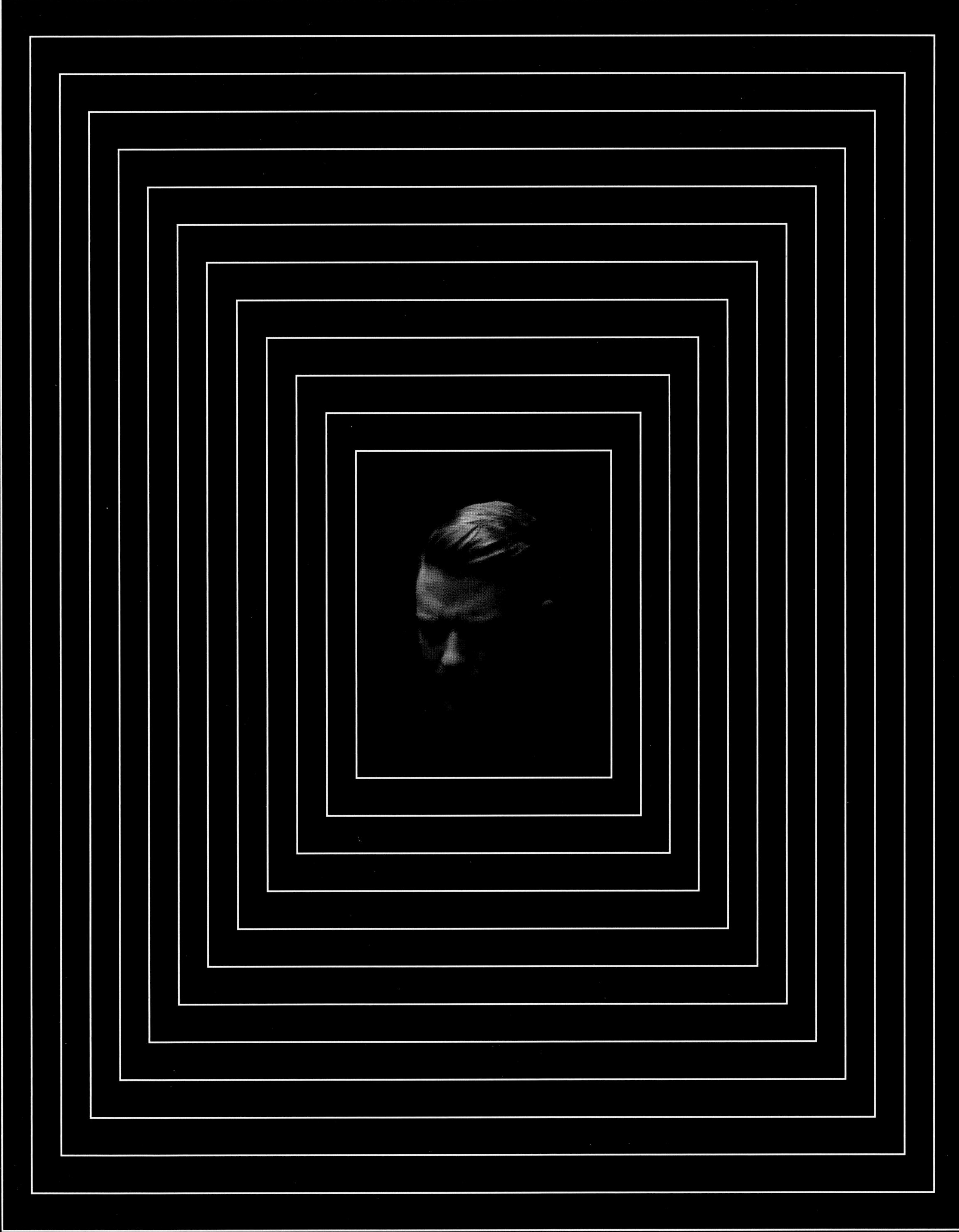

The Tom Ford of It All

LIFE AND ART SOMETIMES MELD IN WAYS YOU DON'T EXPECT. Just after I finished writing and recording *20/20*, and just before the tour, I was planning my wedding. I called Tom Ford. I knew I was getting married only one time in my life, and I wanted to look a certain way. Tom agreed to dress me and my groomsmen, which ended up being his wedding gift to me. He just did it all for us. That's the type of person he is. Then came the tour. When we were planning the show, everything still felt like a Tom Ford world to me. That was the world I wanted to live in then. Working with him was incredible. *20/20* was the first tour that I did with that group of people. There were so many of us, and Tom Ford dressed us all: me, the dancers, the Tennessee Kids, the whole band. And if you saw that show, you know there was a lot of band. It was an entire vision, beginning with sketches, that became the aesthetic for the whole *20/20* movement. It felt right and we went with it. All the clothing has a very specific tailored silhouette—inspired by the days of the Rat Pack and a golden age in Hollywood—and the attention to detail that is Tom Ford's signature. The clothes captured the vibe of everything I was doing on that record. To me, the band looked just the way the music sounded. "Suit and Tie" was one of the last songs written for that album. The music inspires the vision, and the vision inspires the music.

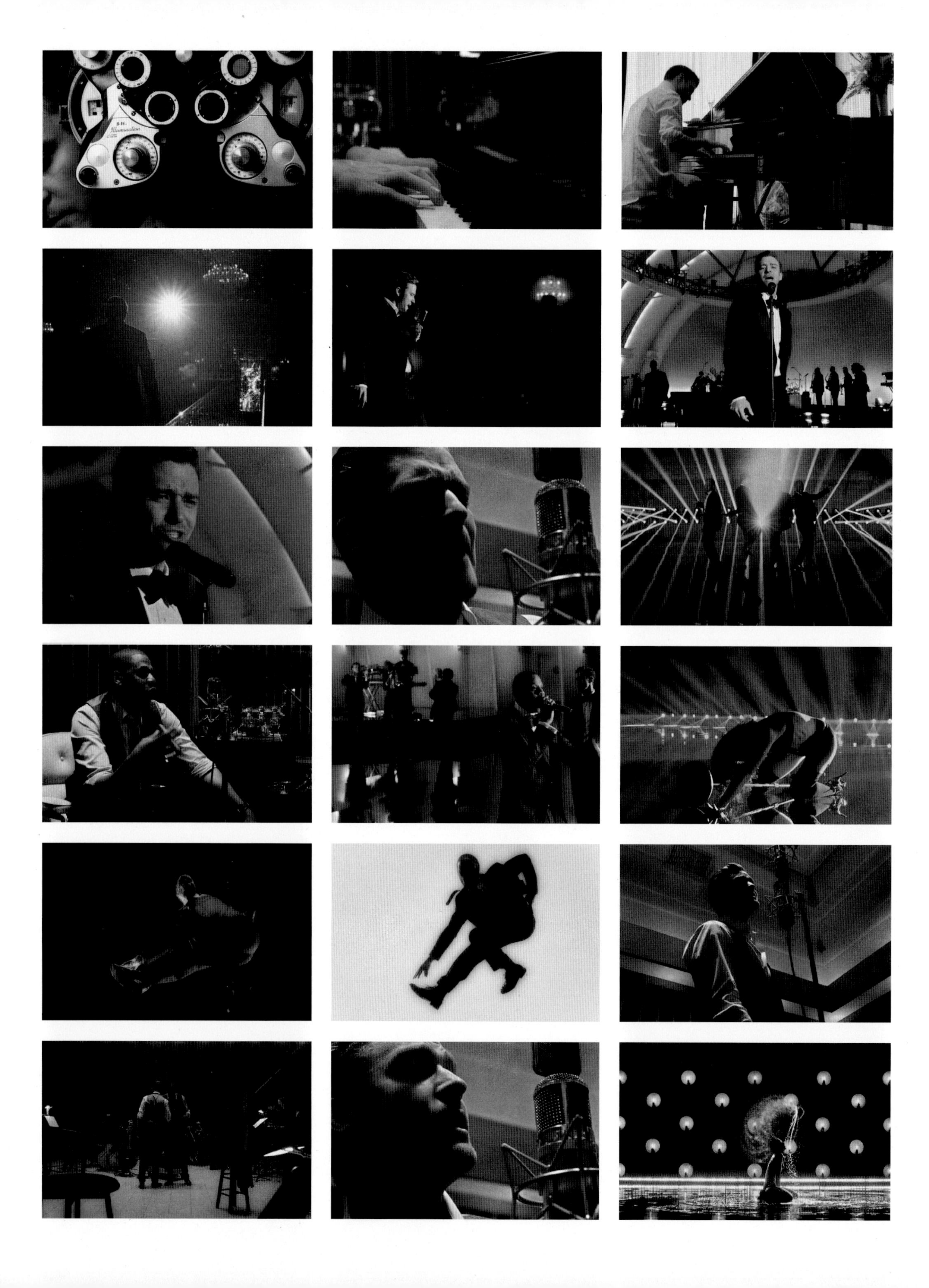

PARIS

FedEx
Borsalino
NIKE

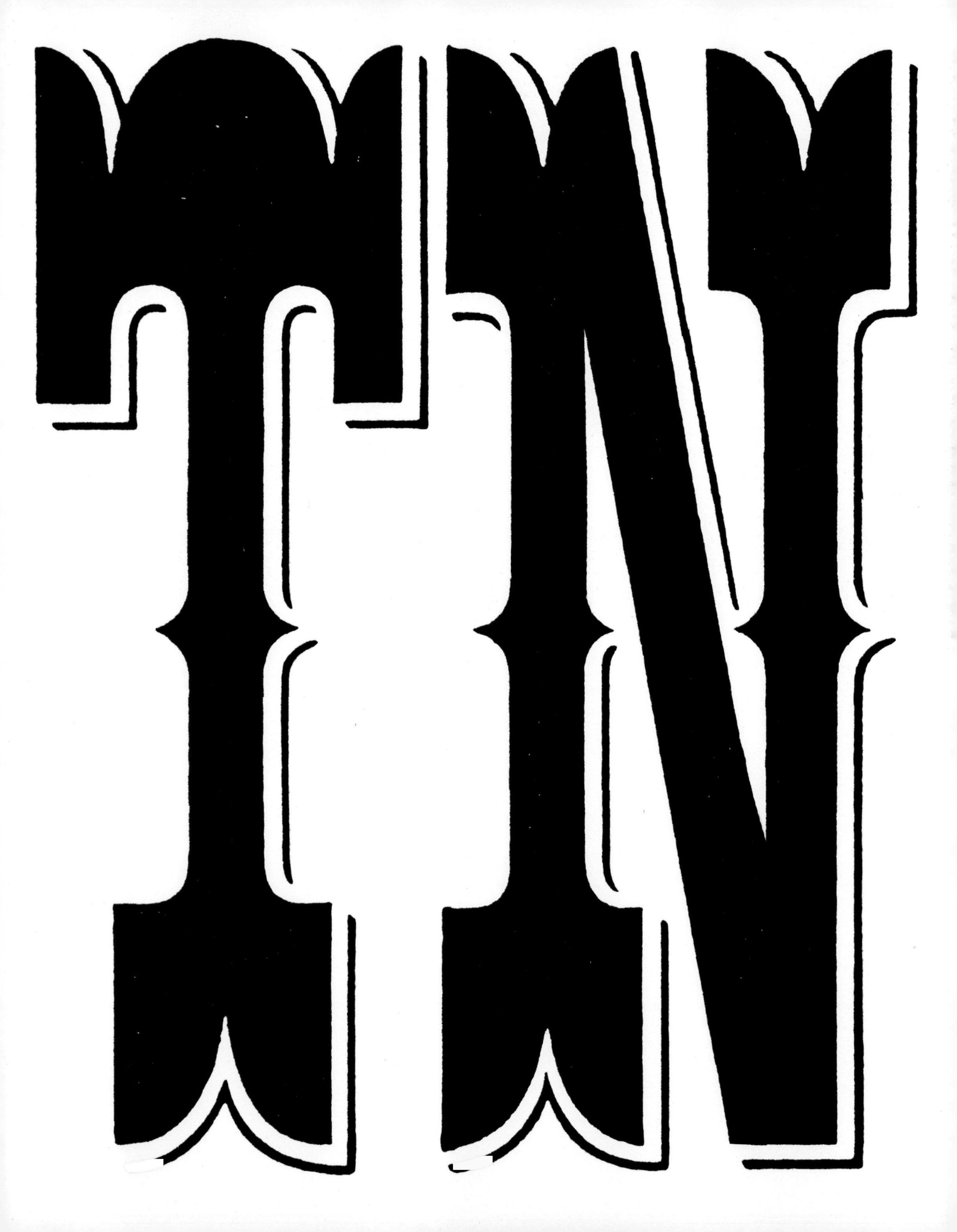
TN

The Legend of the Tennessee Kids

FUNNY ENOUGH, THE TENNESSEE KIDS STARTED WITH ME JOKING around. I sang background on a song on *The 20/20 Experience* called "That Girl." And I sang the backgrounds differently than I normally would have. I sang them like I was pretending I was in the Four Tops or the Temptations. I wanted to name the guys, who were all characters in my mind as I was singing. I named them the Tennessee Kids, and that's how I credited them in the liner notes: "Background vocals by the Tennessee Kids."

When it was time to go on tour, I said, "Well, we'll just be JT and the Tennessee Kids." Bruce Springsteen and the E Street Band was a big inspiration for that.

I recruited some of the most badass musicians and dancers in the world, some of whom I'd worked with before on the road and in the studio or made videos with, and even a few who I had just recently encountered before *The 20/20 Experience* world tour. They were from all different walks of life, all different places, California, Georgia, Pennsylvania, Texas, even as far as Russia. The feeling I get playing with these people onstage is surreal. There's something that happens between artists who have a mutual respect for each other that somehow makes you symbiotic. When we play together, we really play together.

The Tennessee Kids refers to the band and the dancers, but, really, it's inclusive. We're a family. The people who come to the shows, everyone who participates, who listens, who feels something in response to the music—you're all Tennessee Kids.

To me, anybody who's involved is a Tennessee Kid. It's a movement. I wanted to make everyone feel welcome to join us on the journey.

adidas

My Friend Jonathan

THE 20/20 EXPERIENCE TOUR WAS MONUMENTAL FOR ME AS AN ARTIST. ABOUT halfway through what was a year-and-a-half-long tour, everyone, from my managers to my best friend, kept saying we should really capture the show on film because it was such a unique thing. I believed that people were responding to the tour this way purely because of all those incredibly talented musicians and dancers who were onstage with me: the Tennessee Kids.

There was only one auteur who I thought would be perfect to make that film: Jonathan Demme, who directed *The Silence of the Lambs*, which he followed up with *Philadelphia* just two years later. That's the best one-two punch ever, maybe. The concert film that Jonathan directed for the Talking Heads, *Stop Making Sense*, is still the best concert film ever made.

When Jonathan agreed to direct the film, he came to see one of the shows. I remember his response. He said, "You know how concert films always cut back to the audience response? We're not going to do that here. All the magic is happening onstage. There's a real electricity between all of you and what you're creating up there on that stage every night. That's the show."

He then put his film crew together and came to see five or six different shows to really zone in on where he wanted the cameras and what he wanted to capture. He saw what I knew to be true: that the show wasn't just JT, it was JT and the Tennessee Kids.

He shot the concert in Las Vegas on New Year's Day and on January 2, 2015.

He passed away just months after it aired. I still think of him often and miss him very much. I'll never forget him. He was a special person.

He would call me up randomly and say, "Hey, remember that conversation we had? I just wondered how you were doing."

He was one of the most beautiful people I've ever met.

JT

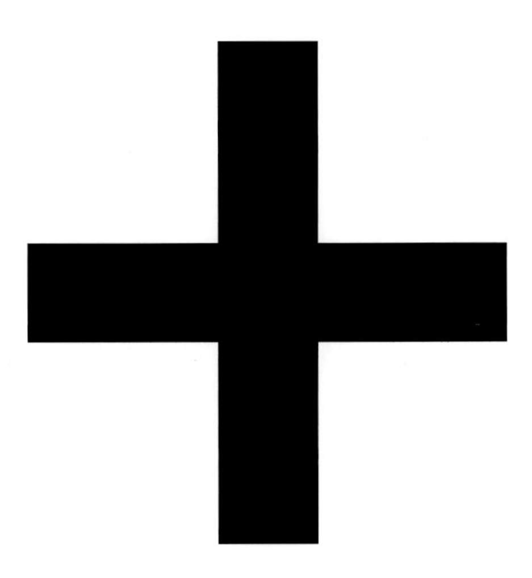

I WAS ON *THE 20/20 EXPERIENCE* TOUR when my life changed forever. I was in Detroit and pulling a week or two by myself because my wife was working in Los Angeles, but was coming to visit me in a few days. I had just finished a show, and I got into my car and took my phone out. There was a text from Jess, in all capital letters. It read "CALL ME NOW." I FaceTimed her back immediately. She answered with this look on her face, and I knew exactly what she was going to say. Instead, she just held up the pregnancy test, and we both started bawling. I ended the tour early so I could be at home and take care of my family. My last show was in Las Vegas, on January 2, 2015.

THREE MONTHS LATER, I WAS A FATHER.

The Stillness

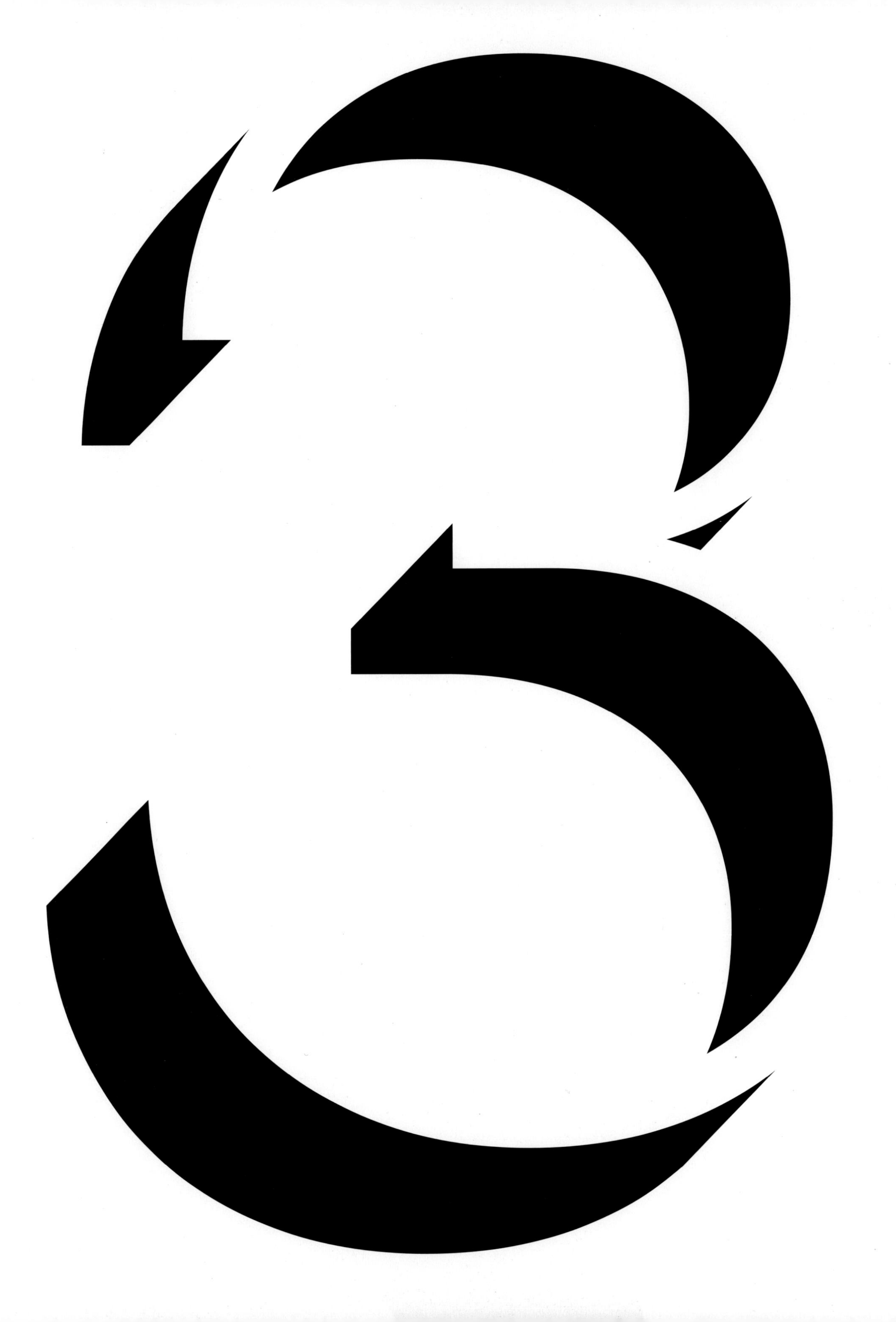

“So the darkness shall be the light, and the stillness the dancing.”

T. S. Eliot
“East Coker,” *Four Quartets*, 1943

Life is movement. If I want to find my stillness, I have to find it within the noise. Noise is like breath. It's always there. But even the sounds of a busy city like New York can become white noise, almost invisible, if I let it pass over me. If I want to quiet the noise, I find the quiet in the noise. True stillness doesn't come from an external place. It lives within me. It's that feeling when I can see how beautiful everything really is, even when it is painful. When I understand that the dark and the light are all part of the same story. That kind of stillness lets me see who I am. What I have been given. Who I might become. And what I have to give. Finding your stillness is about becoming yourself, just continually becoming.

"If you had

ONLY ONE WORD TO DESCRIBE YOURSELF, WHAT would it be?"

I've been asked that question more times than I can count. Can you answer it? I can't. Because we're never one thing.

That's confusing for some people.

What confuses me is why they feel like I need a definition.

Why do I need to be defined by anybody else? Why do you? I can't be defined by anyone but me. I can tell you who I am or what I am, but not if you want me to be one thing. Because the only thing I'm sure of is that I'm not just one thing, nor will I ever be. That's an idea that continually excites me. I would like to stay in a place where I don't know exactly what I am, or what I am doing, because I would get to continually discover myself within that.

I'm always in transition. So are you. So is this whole planet. It only appears like the sun is moving. Remember, we are the ones spinning around in space. You're always moving. But sometimes it feels like everybody wants you to stay right where you are. Everybody's always looking for a definition, a classification, a rule. They want to pin you down so they can understand you. They want you to make it easy for them. They want you to walk in a straight line.

I say, walk your own line.

Once I was at a songwriting conference with Bill Withers, one of my heroes. He reminds me so much of my grandfather, in that they are both the kind of man who will tell it like it is, who will always be straight with you.

Bill has a daughter who is a musician. In the documentary *Still Bill*, you see her working on her music, while Bill is talking about what he tells his children about the creative life. He says, "It's okay to head out for wonderful. But on your way to wonderful, you're gonna have to pass through all right. And when you get to all right, take a good look around, and get used to it, because that may be as far as you're gonna go."

Then there's a moment where he finally hears his daughter's song, a song she'd been working hard on and holding on to for a long time. He hears this song, and he just burst into tears. That's the kind of man I want to be.

Bill told me once that people don't know or care who you are. They don't take the time out to know who you are. They are interested only in what you are.

He told me that to talk about who you are now, you have to find out who you were before, so you can understand *why* you are.

Younger aspiring artists sometimes ask me for a blueprint. I tell them what I will tell my son one day: Don't do it like I did. *Do it like you do.*

Don't do it
like I did.

Do it like you do.

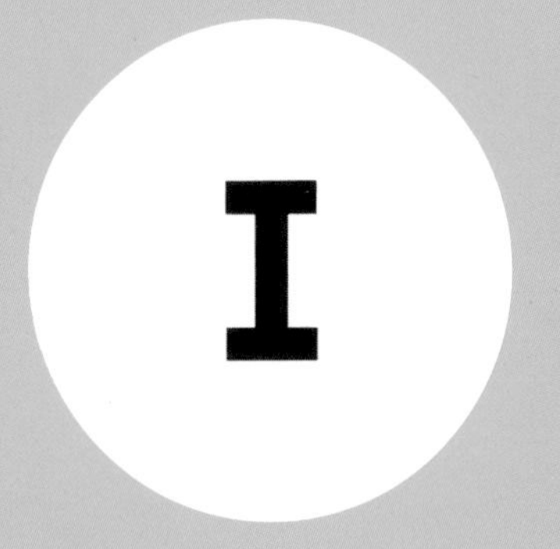

I STARTED PLAYING GOLF to get outside. To have a reason to walk around in the grass for a few hours, to quiet the noise of my thoughts, to learn how to stand in front of this small, still ball and figure out how to shift my weight and swing my arms to send it sailing up and over toward the trees in the distance, or coax it to move slow and sure with the break of the green and into the hole.

When I was twenty-one, I was on my first solo tour. I was touring for *Justified* and I was at the point where I felt a little insane. I was just going from arena to arena. We would play a venue, and I'd realize that I'd already been there. When you're touring and playing arenas, you're indoors all the time. You're always in darkness, never in the sun. The shows are late, so you try to sleep in. You get up around eleven, exhausted from the night before, but you have another show that day. You eat a late breakfast, get your body moving in the gym—and you're still indoors. You drive from the hotel to the venue. Sound check's at four thirty. And you're in the venue for the rest of the day. Inside.

The whole routine was giving me cabin fever.

I remember my stage manager coaxing me to get outside with some of the crew one day off and come play a few holes. Mind you, my stepdad used to try to get me into golf when I was a kid. I was into it for about two seconds. But really, I wanted to play basketball. I just didn't have the patience. So, picture me out with a crew full of guys, about seven to eight of them, and a few beers later . . . I step up to the eighteenth tee box and blister a drive right down the middle of the fairway. That was the moment that I knew this game would get all the way under my skin.

I started with golf because I wanted to get outside, and then I realized that it can be a social and fun time with your friends, because you get to hang out for a few hours—that's how long it takes to play eighteen holes, more or less. It can also be very meditative and personal, because you have the time and space to just appreciate something beautiful. A golf course is like a park, all fairways and trees. And depending on where you're playing, you might see a little of the local wildlife, a hawk, an owl, a deer, even an alligator. Nature is a powerful thing, and some of the best courses in the world embrace that ethos.

It's also the hardest sport I've ever played. Like, impossible.

Golf is nuanced, it's physics and math, and it requires the kind of close attention that allows me to let go of other things. It's not a reactive sport. The ball is still, and you are still before it, before you decide what the motion will be. No shot that you play in any round will be the same, either—they're all different.

Now I can admit that golf has captured me, and I have fallen in love with my captor. Golf is a sadistic sport. I can never please it. It's fickle. It's elusive, too. The minute you think you've got it figured out, you discover that you don't. The wind affects the ball, the course affects the game, and you affect all of it. You're constantly coming up on a new moment, a new situation. You have to be creative to make a good shot. Sometimes you're behind a tree. (Hey, man, sometimes you just shank it off the tee . . .) So then you have to bend it around that tree. Or, hit it under that tree. Or just hit your ball toward the halfway house and order a beer.

It's a game that you will never conquer, so you have to show respect for it. I think that's what I like about it, too.

Golf is just you and the elements. You have to be able to work with what's around you. You're still the only one who can mess up your game.

Yep, it's a sadistic sport. And, maybe, I'm just a masochist.

Say Goodbye to the Old Me

W

HEN I DID MY FIRST ALBUM, *Justified*, I wanted to make an R & B album. Reviews came in and they called it a pop album. That bothered me for way longer than it should have. I felt like they just didn't get it. I mean, I understood that I was coming from a huge "pop" group, so it was easier to classify me in that category. Nevertheless, it was my first real piece of work that was all my own, and I had a vision for it. I wanted people to get that vision. I was also twenty-one and precious as hell. It's clearly a pop album.

With the next one, I realized that I didn't have to make people get it. I just have to do what I'm doing. I realized that if I was to be considered a "pop" star I could do anything I want. And the album that was birthed out of that realization was *FutureSex/LoveSounds*.

Now I see what both of those albums were about. *Justified* was about acceptance—wanting to be accepted by my own community and wanting my differences to be accepted. *FutureSex/LoveSounds* was inspired by self-acceptance. That creates a different type of hunger.

Instead of waiting for an invitation, something happened inside of me where I said, "Well, you know what? I'm going to throw my own party. You're all more than welcome to join."

The 20/20 Experience was inspired by wanting to see further and deeper. For that matter, so is this book.

Now I've made more music. I'm exactly where I'm supposed to be. Now I fully realize that I can do any style of music that I want because I grew up listening to everything. Sure, it will be my version of that genre, but this is where my own originality meets my influence. If I could give a young artist advice about their influences, it would be this: Embrace them. Don't run from them. They will be there whether you like it or not, so identify them and honor them by being influenced by them.

And then? Just be. Be yourself. You are the common denominator. But you have to go through a lot of things, you have to go through and come back full circle to get to that place.

That's what the album *Man of the Woods* is about: introspection.

Jay-Z told me once that one of his buddies from the Marcy Projects, in Brooklyn, said to him, "Man, you've changed." And he replied, "You're goddamn right. You act like I've been busting my ass to stay the same."

That's how I feel. I busted my ass to get where I am. I have no expectations about what comes next because I feel that I've made something beyond what I wanted to make. It's the carbonation. I couldn't not make it. As an artist, I feel better and clearer than I've ever felt in my whole life.

No matter what gets thrown at you, you catch it. You learn from it. And you figure out how to throw it back.

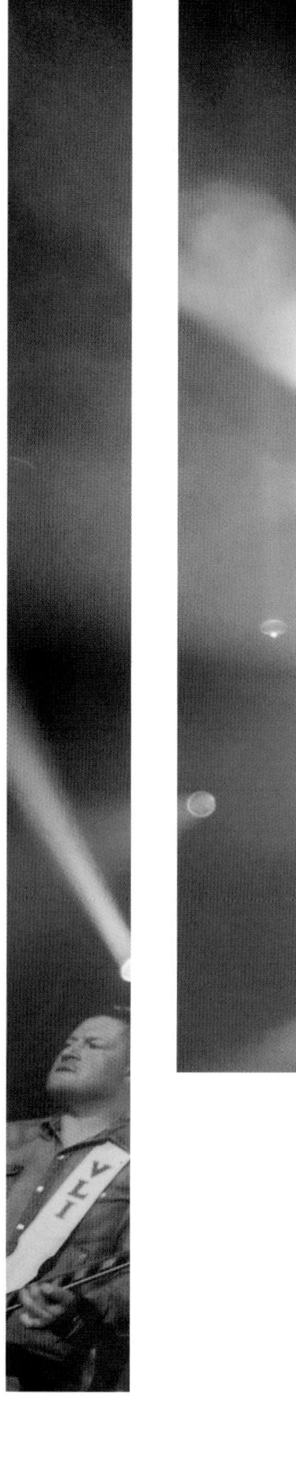

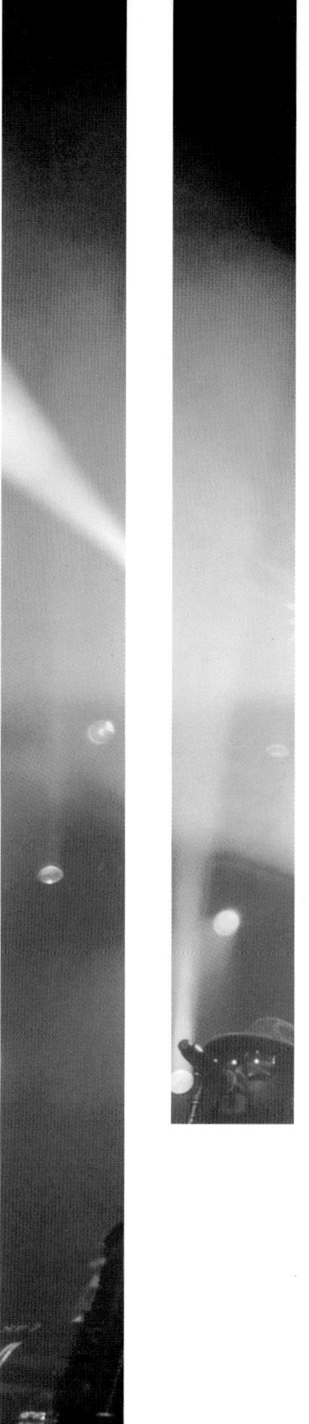

I realized that

I didn't have to

make people get it.

I just have

do what I'm doing.

A Man Becomes a Man

JUST COMING INTO MY OWN, AND I'M AT AN INTERESTING crossroads in my life. I've been getting here my whole life, but now more than ever, I can see all the pathways—they're right in front of my face, and it feels good. I have space to look back, to look forward, to think about what I value, and why.

What I understand now is that we're all just figuring it out as we go. We don't always have time to reflect, which can lead to a kind of tunnel vision. I can see that the first half of my twenties were all programmed toward achievement, but somehow, as I hit twenty-six and twenty-seven and twenty-eight, I was three different people within those years. At twenty-nine, I started to really understand what I'm good at, what I'm not good at, what I need to work on. What has really helped me see that is the birth of my son, Silas.

Having a child has been the pinnacle of my life. The things I have learned from being a father have been profound. It's made me look at my relationships in a new way. It's made me think about who my parents are as people, and how that affected me then and has contributed to who I am now—and how all that might affect my child.

It makes me wonder how my son will see me when he grows up. It makes me want to be more thoughtful.

I've spent my whole career shooting from the hip, living in a certain way, creating in a certain way. In my thirties, I still shoot from the hip, but now when I shoot, I'm aiming.

I've seen the distance between me and my target before. I'm aiming at something specific, too, instead of just running around like a cowboy and throwing stuff against the wall to see what sticks. I'm learning to wait for the right moment.

When my son was born, my wife and I wanted to keep people away from him. We wanted him to be just for us, just for a little while. When we wanted to share him, we thought about how to do that in a way that felt good to us. Something fun. I own a piece of the Memphis Grizzlies, and the basketball playoffs were starting, so we put him in a little Grizzlies outfit and took a photo. That was the first picture that anyone had seen of him.

It was important to us to choose how to share him with the world because this is a whole new era for me. It's no longer just about me. I have a wife, a child—a family. It's terrifying. It's invigorating. It's more meaningful than anything I've ever been a part of. Kids teach you more than you teach them. I have absorbed so much from my son. If you're reading this and you're a parent, you know what I mean. I'm learning something every day. And by something, I mean eighteen things.

Here's one thing I've learned: the real winning at parenthood is getting to wake up every morning and fail over and over again.

That's where I am now. I've changed because I have a kid. Being a dad is an experience where a day can feel like a year, but a week can feel like a day. I've been taking more time to just look around, and I can see how all the small moments are the most special things now. I'm talking about the tiniest things that might have passed me by in the past, like quiet moments at home. When my son smiles at me. When my wife looks at me in a certain way. These are the moments that are so big in my life now.

Being a parent has given me the chance to gain humility. I've worked on discovering my strengths and weaknesses, and to talk about them, to speak up and say, "I'm not particularly good at being patient. So you might have to bear with me on that." (I'm famously impatient.) Or "I'm not good at sharing." (Only-child syndrome?)

With hindsight, I can see that, of course, I've made mistakes along the way. I can see that I didn't really have time to be a kid. I was on the path I was on, and in some ways, it set me back. I've been able to look my childhood more squarely in the eye and see it for what it was.

It takes humility to admit shortcomings, to talk about things. And it's never perfect. It's a practice. Sometimes I don't get it right, but I get to wake up every morning and fail, over and over again.

Wherever I find myself now—it's a good place to be. I find courage in the idea that our mistakes can empower us. They can become the thing that inspires us to make history—our own history. Big or small, the moments we create lead us forward, over a new threshold.

NOW
WHEN I
SHOOT,
I'M AIMING.

This Moment

WHENEVER I WRITE MUSIC, I HAVE TO EXPLORE. I HAVE TO go within myself and see who I am in the moment and what is within me, and invite that feeling to come out into the world and show itself.

The album *Man of the Woods* was inspired by three people: my grandfather, my wife, and my son. The whole album—the lyrics, the sound, and the feeling—was born in a space where I just had a child. There is so much feeling in these songs, and they are so personal—I was having a new awareness of myself as a man, feeling fully realized in that I don't know anything, feeling fully comfortable knowing that there are things I needed to learn, that I still have to work on.

I was writing this album while my wife and I were just getting to know our child, and getting to know each other in this new way. That's why I wanted Jess's voice on the album. She's almost a narrator, or a voice of consciousness, and she even has her own interlude on the album entitled "Hers." It's the way my life feels now—whatever I do, wherever I go, when I come home, it's her voice narrating my life.

She's a different person from when I first met her, but she's also the same. I'm sure she'd say the same thing about me. Ten years is a long time. I used to think we want to be loved for what we are, but maybe more now, I think we want to be loved for what we aren't. I think we want to be loved for all our fucked-up shit.

We should say "I will" instead of "I do" when we get married because people constantly change, both physically and mentally. I've watched my wife change. I've watched her body change. It's a temple. It should be worshipped. It should be marveled at. I'm fascinated by her. She's everything, man. She just constantly surprises me with who she is, and who she's becoming, and I really hope I do the same for her. I'm excited to see what she's going to do next. I wake up and roll over and look at her, and I'm inspired.

That's a special place to live in. That's a special place to write from. It makes me look backward and it makes me look forward, and all the while I'm so in this moment that I can see all of it at once.

When I'm writing, I have to trust in those feelings. I have to trust the process. I have to trust myself. I can't help that my music shows who I am in this moment, what I'm drawn to, what I'm wondering about. I don't want to help it. What you hear in the words, what you feel in those songs—that's what I was feeling when I wrote them.

I want you to see me, just like I want to see you.

That Feeling . . . Inside My Bones

WHEN I WAS YOUNGER, THOSE SUMMERTIME songs were a drug for me. You know what I'm talking about. Imagine your favorite summer memory. Sun out, sunroof down, hair blowing in the wind. Maybe that first cold sip of a lemonade on a front porch. That first plunge into a pool that gives your whole body goose bumps. The five minutes after you come out of the pool and you can feel yourself warming in the sun. There's such joy in those moments. And they hit all your senses. "Can't Stop the Feeling" is my shot at the ultimate summer song as a songwriter. I wrote it to make people feel great, to connect with the best, brightest energy inside themselves. In the end, all the sound, what is added, what is taken away—is all there so that I can look around the world and watch it connect people because it makes them feel good. When I first saw my son, I had such intense feelings—nothing that I've ever experienced can compare to it. I could never have written "Can't Stop the Feeling" without knowing my son. If you want to know how I felt when I first saw Silas, listen to "Can't Stop the Feeling." He inspired me to write that song.

He is my pure joy.

I WAS ONE OF THOSE KIDS WHO WAS ALWAYS BANGING ON THE LUNCH TABLE,

ON ANYTHING AROUND ME, REALLY. I'M NOT SURE, BUT I would assume that the other kids and the adults didn't always appreciate my impromptu concerts. From an early age, I've been curious about and fascinated by the way things sound, and I could hear the subtle shifts in tone, too.

Everything sounds different, has a specific sound. If you tap the table in front of you, the chair you're sitting on, or the window next to you, you hear three different sounds. Richer, fuller, and hollower. I'm doing it as I write. This echoes. And this surface makes a sound that hangs flat. And this sounds more like marble, this more like tin, this more like velvet. All these vibrations. The air conditioner sound. The ice in the glass. *Clink, clink*. That woman walking across the lobby in heels. *Click, clack*.

They each have a sound, and I hear them. All the layers, coming together to make a whole atmosphere of sound. It isn't just in music. It's around us, all the time. All you need to do to hear the music in anything is listen. Just be still and listen.

Walk around a city. The axle in a car makes a rhythm. The hustle and bustle of footsteps on the pavement makes a shuffle. There are sounds in the background as much as in the foreground. I don't know how to listen any other way. The rhythms are everywhere. The possibilities are everywhere.

When we were recording "The Hard Stuff" for *Man of the Woods*, there was a sound I wanted, like fingers drumming lightly on a surface. I looked around to see what could deliver that sound, and my eye fell on a regular leather-covered-seat barstool that was sitting in the recording booth. I went into the room where the vocal mic was set up, pulled the stool up to the mic, and I tapped the sound out on the seat.

We recorded it until I had the right rhythm, and then we rolled it into the song. If you listen carefully to the percussion, you can hear it, looped into the whole beat, acting like a kick drum. This little beat comes from my human hand beating on a surface, just like when I used to beat on the lunch table at school. What I was doing may have sounded like noise to other people, but when it is applied in the right way and it feels good, it's music.

Gibraltar

Sometimes the Greatest Way to Say Something

HE ENTIRE ALBUM *Man of the Woods* only has four features: my wife, whose voice is a constant companion and a consciousness throughout the album; my son, whose voice is at the beginning and end of "Young Man," the song I wrote for him; Alicia Keys, whose voice in "Morning Light" sounds like a warm blanket; and Chris Stapleton.

When I met Chris, I just knew right away that I would know him until I die. Sometimes you know—by which I mean that deep inside your heart, where it matters, you know as much as you've ever known anything that has turned out to be a true thing—that you've just encountered a special person. That's how I feel about Chris.

That kind of connection is what's underneath the music we played together at the Country Music Awards in 2015, which I'm pretty sure took everybody by surprise because I'm not someone you'd expect in the lineup at the Country Music Awards, and Chris was coming off his biggest moment so far in his career with his debut solo album, *Traveller*.

Chris is a prolific songwriter and had written a ton of hits for other artists. This was his time to shine on his own, but when he called me to be a part of it all, I didn't hesitate. We rehearsed a couple of days before, but we also decided to keep the performance loose, not too polished.

First, we did Chris's version of "Tennessee Whiskey," and then an original of mine, "Drink You Away."

That's the thing about working with great musicians. You don't have to overcook it. I realized that years earlier with my impromptu Grammy performance with Al Green. What happens onstage will be great because you trust each other's instincts. And then? Well, the songs with Chris were a special performance on a special night, where he went on to win three major awards, including Album of the Year. I still find myself getting into conversations with grown men about that performance.

When it came time to do *Man of the Woods*, I knew I wanted Chris on the record. We sing a song called "Say Something," which we wrote together. I love "Say Something." The song was spawned by a conversation we had in the studio, and it became the basis for the lyrics. It's basically a confessional or testimonial that as artists we feel a subconscious responsibility to evoke and lead a thought all the way down the path. But then you always have this moment where you feel like you're looking in a mirror and saying, what the hell do I know? I'm not the person to be telling someone else what they should be doing or not doing. I don't know what I'm doing—which led me to what I think is the greatest line of the song—"Sometimes the greatest way to say something is to say nothing at all."

While I was mixing the album in New York City, Chris came down to listen to the way the tracks had turned out. I had been in the studio for weeks, fine-tuning every sound. That's how I work. Beat by beat, moment by moment, line by line, track by track, until it all hangs right to my ears. But it isn't easy to get there. It's a lot of trying and listening and trying again and listening harder, closer, and sometimes, listening from farther away. Recorded onto a phone and played back without the benefit of a studio-grade sound system. Or even farther away, out of the studio altogether, in a car, which is another good way to hear what you've been doing.

The tracks weren't all the way finished, but close enough for me to play them for Chris, to show him the magic we had created together.

He was there for the dreaming of the songs, and he could understand the relationship between the tracks we laid down, just the two of us singing with each other, playing our guitars, and what I had taken and layered on top to make it into what you hear it as when it's finished. And God, it's beautiful.

We wrote three songs together for this album, and we wrote many more songs together that are sitting in the vault somewhere. Maybe they'll see the light of day someday.

Chemical Reactions

EVERYBODY brings what they bring into the room, and I bring what I bring, and what you get out of the mix isn't always what anyone expects. I worked with Pharrell Williams on my first record, and I worked with him again on *Man of the Woods*. To come back together after all this time was an interesting experience. We got to spend a lot of time on the record, took walks out in the grass, and talked about life. And all of that spilled over into a lot of the lyrics.

Pharrell and I are very different people, but we have a similar chemistry. We have a lot of fun together. He's a wonderful collaborator and he's always got a very strong point of view. He mixes quirk with musicality, which is very cool. He comes at things with an idea of being different, and I like that, because I always want to do something new as well.

JT

BROOKLYN
MACHINE
WORKS

Then and Now. When I'm thinking about who to work with, there are all these algorithms I have to consider. When you're producing an album, you're doing a lot of different things at the same time. You play with a groove until it feels just right. You sit down and write the chords for the bridge on a certain song. You hum and craft a melody until it sits on top of that chord progression and grooves perfectly at the same time. You have to be malleable in different ways. And you have to step back and admire what the other people who are working with you are doing. I don't have any ego about whatever's best with whomever I'm collaborating with. I worked on the bulk of the album with Pharrell, but I also wrote and produced "Filthy" with Nate "Danja" Hills and Timbaland, my main collaborators on *FutureSex/LoveSounds*. Along with "Say Something," Chris Stapleton and I also wrote "Morning Light." That song I produced with Rob Knox, who has been a collaborator with me for a while on songs like T.I.'s "Dead and Gone" and numerous tracks on *The 20/20 Experience*. I feel very comfortable with these guys, and we all trust one another. When you have that level of trust, you can go outside of your comfort zone and know that you can fall back on each another to be honest and find the magic within a song. Whatever it calls for, whenever it calls for it. **Be still and listen.**

You have to be
malleable in

malleable in
different

You have to
malleable in
different ways

Into the Dark Forest of Memory

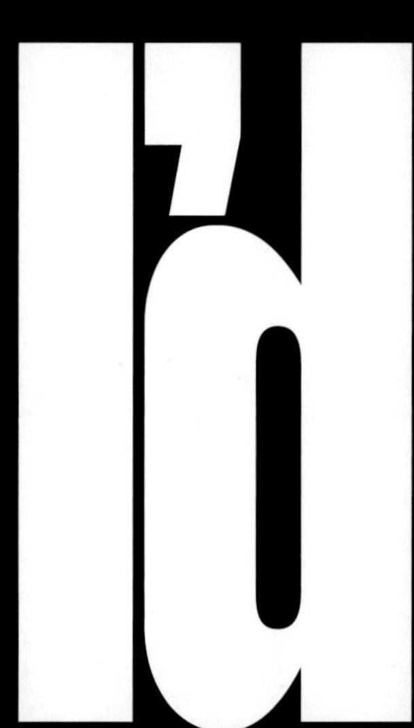

LOVE TO GET TO THE BOTTOM of how I feel about everything.

When my son was a baby, he was so clear with us about what he needed. It was an uncensored, completely innocent expression of what he wanted, and it came out as a tirade. We call it a tantrum, but he was just expressing himself.

He wants what he wants. He doesn't bury it.

The unconscious is a weird place. It's where you find your history. Your experiences. Your fears. Your feelings. Your trauma. What you want, what you like, and what you don't like. Your desires live there and your prejudices do, too. The unconscious mind is also where you find your compassion. We need to find that compassion to reprogram our prejudice.

I see my son looking at me sometimes, when he doesn't get what he wants, and it's so pure. I see him looking at me and thinking, *You're not giving me what I want, so you must not be listening to me, which means you must not love me.*

My wife and I make sure to tell him that we hear him. That we know he's upset. And that it's real. It's okay to be upset and not just pretend to be happy. It's okay to admit that you want things.

That's something I had to learn. It's not something I was taught.

I don't know what my grandfather thought. I don't know what he felt. He didn't share that with me. That's why I wrote "Livin' Off the Land." It was for my grandfather, so I could try and understand his world. It's a song about making it on your own, and on your own terms, out in the natural world. It's not the world I live in, but it was the world of one of the most important men who brought me forward.

While we were in the studio, we included a few clips from the show *Mountain Men* in the song. The clips had just the feeling I wanted. You can hear them at the beginning of the song, these low, gravelly, intense voices plainly explaining their relationship with the land. You know that those voices are real and have seen real things. Connecting to that feeling, trying to imagine it, to climb inside it—it's all a part of looking deeper into myself: where I came from, and who raised me, and who raised them, and what they were taught, and what they taught me, and what I will teach my son.

My son sometimes wants his mom; he just doesn't want me. I can't give him what he needs, sometimes, and he pushes me away. I'll feel bad for a moment. I'll feel inept. Why can't I help him? Why doesn't he love me? I have to remind myself that of course he loves me. But she's his mother, and that's who he wants right now.

For a long time, I thought I was good about voicing my feelings. If I had an issue, I said something right away. That's what I thought, anyway. But as I get older, I realize that I have to be more honest with myself about what is really going on. Even though I was always good at a very moment-by-moment, visceral way of telling someone when I was uncomfortable, there were other things, deeper-seated things, that I'd never talked about.

Until I became a father, I thought I had things to be afraid of. Now I understand that I don't have to conquer my fears. I just have to learn to live with them. Instead of looking for sunlight to erase that shadow, I know that it's always going to be there. Everything you're doing as an adult is to try to heal whatever you've built up from your childhood. It's your shadow that follows you around, and the only time a shadow disappears is when you step into the darkness. And then you learn to step back out.

Now I can finally see the relationship between all of it, and it's beautiful. In all of its ugliness, it's beautiful.

Understanding that allows me the stillness.

Arms Open

Outro

I like the silence. I love being alone in my car, holding the wheel. I drive with the windows up, and I rarely play music.

I crave stillness. I know that if I want to embrace the feeling of stillness, I have to find it inside myself. It began as a need, and then it became a practice. Finding the pieces of myself that I can share with other people is also part of it; that's connection.

This is my cycle. This is who I am as a man and an artist. I imagine a life where I don't know where I'm going. I just know that I am. What a journey that will be. The waves come. I let them come. And I know that I'm rooted enough to withstand any of them.

It takes a lot to trust being still, but when I do, I see how powerful stillness is. The best moments I can remember are not when I reached out and grabbed for those things. When I've given the best I've got, I don't have to make a lot of noise. I simply let it come to me.

All I have to do is open my arms.

Acknowledgments

From the beginning, Rachael Yarbrough has been my champion—always there to lend a trusted ear and eye, with all her heart.

Rachael, we have known each other through eras. We have seen one another through so many changes and so much growth. I will always appreciate what you have given me and what you have taught me, and I will always believe in you as much as you have believed in me.

I couldn't have written this piece of my life and made it so meaningful without the help of my great friend and "producer" on this project, Sandra Bark.

Sandra, you are a beautiful soul, and I am in awe of your gift. I hope you always look back at this book and the time we shared together with as much reverence as I do.

I forever will.

References

Hiatt, Brian. "A Final Visit with Prince: *Rolling Stone*'s Lost Interview." Rollingstone.com, May 2, 2016. http://www.rollingstone.com/music/features/a-final-visit-with-prince-rolling-stones-lost-cover-story-20160502. Accessed August 11, 2017.

Eliot, T. S. "East Coker," in *Four Quartets*. New York: Harcourt Brace, 1943.

Rose, Judd. "Encore: A look at the many faces of David Bowie." CNN.com, September 29, 1998. http://edition.cnn.com/SHOWBIZ/Music/9809/29/david.bowie/. Accessed October 16, 2017.

Rumi, Jalāl al-Din. "Desire and the Importance of Failing" in *Feeling the Shoulder of the Lion*. Translated by Coleman Barks. Boston: Shambhala, 1991.

Photography Credits

Alamy: 21 (left): Nikreates/Alamy Stock Photo.

AP Photo: 226 (left).

Bishop, Lloyd. 172: © 2013 NBCUniversal Media, LLC.

Disney Channel: 37.

Everett Collection: 39, 43: © Buena Vista Pictures/Courtesy Everett Collection; 86, 87: Merrick Morton/© Columbia Pictures/Courtesy Everett Collection.

Fallon, Jimmy: 165.

Fluke, Becky: 256–259.

Getty Images: 8: Erik Tanner/Contour/Getty Images; 20 (left): Andrea Morales/*Bloomberg*/Getty Images; 44: Bob Berg/Getty Images; 46: Nuts Heinacker/Ullstein Bild/Getty Images; 47 (top): Purschke/Ullstein Bild/Getty Images; 47 (center): Todd Maisel/*New York Daily News* Archive/Getty Images; 47 (bottom) and 88: Larry Busacca/WireImage/Getty Images; 48 and 162–163: Kevin Mazur/WireImage/Getty Images; 49, 92: Jeff Kravitz/FilmMagic for MTV/Getty Images; 50–61, 71, 93, 171: Dana Edelson/NBC/NBCU Photo Bank/Getty Images. 91: James Keivom/*New York Daily News* Archive/Getty Images; 166–167: Lloyd Bishop/NBC/NBCU Photo Bank/Getty Images; 174–175: Douglas Gorenstein/NBC/NBCU Photo Bank/Getty Images; 226 (right): Chris Ryan/Corbis/Getty Images (3); 262–263: Jerod Harris/WireImage/Getty Images.

Harless, Lynn: 22–25.

Jarvis, Devon: 31.

McGinley, Ryan: 156–157; 159; 210–211; 216; 234–235; 236–237; 238–239; 241; 278–279; 283.

Netflix: 202–204: Jonathan Demme.

Nguyen, Mark © Tennman Entertainment, Inc.: 12–13; 76–79; 82–85; 94–95; 102–113; 124; 128–131; 141; 144–148; 186–189; 197–199; 212–213; 220; 221; 228–231; 242–243; 251; 254; 267 (bottom); 268 (top); 270–271; 276–277.

RCA Records: 180; 181; 184–185.

Shutterstock: 42–43 (background): Login/Shutterstock; 63: Mega Pixel/Shutterstock; 160: IKasparus/Shutterstock.

Universal Television, LLC: 64; 72–73.

Urbano, John © Tennman Entertainment, Inc.: 178–179.

Yarbrough, Rachael © Tennman Entertainment, Inc.: 16–19; 20 (right); 21 (right); 29; 66–69; 80; 81; 88; 96–99; 120–123; 128–129; 132–134; 142; 143; 149–153; 155; 177; 200–201; 214–215; 218; 219; 224–225; 227; 232; 245–250; 260; 261; 264–265; 267 (top); 268 (bottom); 272–273.

Young, Faith-Ann © Tennman Entertainment, Inc.: 74–75; 100–101; 114–115; 126–127; 154; 182; 191; 192–196; 206; 252–253; 266; 268 (center).

Hindsight

HarperCollins books may be purchased for educational, business, or sales promotional use. For information, please e-mail the Special Markets Department at SPsales@harpercollins.com.

First published in 2018 by
Harper Design
An Imprint of HarperCollins*Publishers*
195 Broadway
New York, NY 10007
Tel: (212) 207-7000
Fax: (855) 746-6023
harperdesign@harpercollins.com
www.hc.com

Distributed throughout the world by
HarperCollins*Publishers*
195 Broadway
New York, NY 10007

ISBN 978-0-06-244830-9
ISBN 978-0-06-288740-5 (BAM)
ISBN 978-0-06-288705-4 (Barnes & Noble)
ISBN 978-0-06-289095-5 (Indigo)
ISBN 978-0-06-288723-8 (Target)
ISBN 978-0-06- 288724-5 (Walmart)

Library of Congress Control Number: 2016939868

Book design by Michael Bierut and Britt Cobb, Pentagram

Printed in China
First Printing, 2018

About the Author

JUSTIN TIMBERLAKE is one of today's most successful entertainers, having earned ten Grammy and four Emmy Awards. As an actor, he has starred in acclaimed films such as the Academy Award–nominated *The Social Network* as well as box-office hits including *Friends with Benefits*, *Bad Teacher*, *Trouble with the Curve*, and *Inside Llewyn Davis*. *Time* named Timberlake one of the 100 most influential people in the world in 2007 and 2013. His 2013 album *The 20/20 Experience—The Complete Experience* was the bestselling album of the year. In 2018, he released *Man of the Woods*, his fourth number-one album on the *Billboard 200* chart.